Composting

Start Your Composting With the Ultimate Eco-friendly and Budget Friendly Techniques

(How to Create Natural Fertilizer at Home)

Scott Johns

Published By **Simon Dough**

Scott Johns

Composting: Start Your Composting With The Ultimate Eco-friendly And Budget Friendly Techniques (How To Create Natural Fertilizer At Home)

ISBN   978-1-998901-09-8

No part of this guidebook shall be reproduced in any form without permission in writing from the publisher except in the case of brief quotations embodied in critical articles or reviews.

Legal & Disclaimer

The information contained in this ebook is not designed to replace or take the place of any form of medicine or professional medical advice. The information in this ebook has been provided for educational & entertainment purposes only.

The information contained in this book has been compiled from sources deemed reliable, and it is accurate to the best of the Author's knowledge; however, the Author cannot guarantee its accuracy and validity and cannot be held liable for any errors or omissions. Changes are periodically made to this book. You must consult your doctor or get professional medical advice before using any of the suggested remedies, techniques, or information in this book.

Table Of Contents

Chapter 1: Composting Recipes

Now which you apprehend the significance of the right carbon to nitrogen ratio, you could start following a few "recipes" for composting. A properly recipe, when accompanied carefully, can make the distinction between first rate compost and a stinky one. In truth, many composters will examine making compost to a little bit like baking a cake. You will add the proper components, in the precise quantities, you will stir your mixture after which you will let it "bake" for a sure quantity of time.

"Baking" Kitchen Compost

There are masses of things that you use for your kitchen on a daily foundation as a way to provide you with an excellent foundation for your composting. You may even be amazed at the stuff you typically throw away that you could be recycling into your compost. You can use a aggregate of some or all the following whilst you operate kitchen components on your compost:

Vegetable peels and vegetable seeds

Eggshells

Fruit peels and fruit seeds

Nut shells (peanut shells are mainly useful)

Coffee grounds (you can additionally use espresso filters)

Fruit and vegetable scraps

Baking Yard Compost

Again, the identical main applies to using your backyard waste to your composting. You can use all or some of those "substances" in your compost pile.

Hay and straw

Wood chips, untreated

Grass clippings

Weeds and different garden waste

Leaves

Manure

Sawdust

Shredded paper and cardboard

Other Ingredients to Compost

There are first-rate deals of different "things" you could compost. These are all elements you likely have already got around your own home. The following low preservation compost recipe will yield outcomes which can be powerful, but it is slower to make. Depending in your ratios and the temperature, it may take months for this recipe to "bake". You will need:

Straw

Dried leaves

Untreated sawdust in small amounts

Untreated timber chips in small quantities

Cardboard portions

Shredded newspaper

Dryer lint

Pine cones

Egg Shells

Corn cobs and corn stalks

Shredded brown paper (such as brown craft paper or grocery bags)

Pine needles

This recipe will come up with rapid results and you will discover that the compost might be equipped in only some short weeks. You will want:

Manure (cow, horse, rabbit, goat, fowl)

Fish meal

Blood meal

Sea weed

Coffee grounds

Lake moss

Cottonseed meal

Alfalfa and/or pea clover

Green lawn waste, which includes weeds

Hair

Kitchen scraps, in particular the ones from vegetables

Grass clippings from untreated grass

Sod

Ingredients to Avoid in Your "Recipes"

While there are numerous matters that you could add in your compost, there are things that you ought to absolutely keep away from for a variety of motives. Here is what you must constantly keep away from adding to your composting pile.

Meat merchandise, bones, fish, dairy merchandise and oil: All of those will motive odors and will virtually appeal to no longer best bugs however rodents as nicely. These items also do now not damage down as properly or as speedy as different organic rely.

Pet feces: You already recognise that some animal droppings are first-rate additions on your compost. Horse, cow and hen manure works very well. However, puppy feces must by no means be used. Cat and canine feces do contain sickness and could damage your composting efforts.

Weeds with seeds and runners: When you upload in seeds and runners for your compost pile, wager what? It will grow and spread to your compost. You can upload weeds, but make certain they may be either useless, or don't have any seeds or runners.

Insect inflamed plants and flora which are diseased: While it is able to be very tempting to throw in those demise houseplants or other flowers with small insects, this will reason issues. These sicknesses and bugs will no longer die for your compost. In fact, they'll spread.

Treated wood shavings and sawdust: Any form of handled wooden have to be averted. This is because it consists of pollution and robust preservatives so that it will damage your compost. If you intend on the usage of wooden, make certain that you only use timber that is untreated.

Ashes: Using an abundance of ashes for your composting will motive the composting technique to gradual. It is normally better to keep away from ashes.

Other matters that you have to keep away from on the way to get the best effects encompass persistent weeds which include poison ivy, bindweed, quackgrass and multiflora. Human feces have to additionally be averted. Any flowers that have long past to seed ought to additionally be avoided.

Mix and Stir

When you have all of your components for your compost, you may need to mix and stir. Composting does not paintings well in case you just allow it sit outdoor. Again, like a cake or cookie recipe, it works better while you add the components within the proper order and also you comply with the directions cautiously. You recognise what happens when you don't comply with a kitchen recipe efficiently. The identical issue can occur with compost. You will turn out to be with a pungent mess. Composting is straightforward, however best while you take some time to learn how to follow the "recipes" of correct composting.

Steps to Mixing and Stirring Your Compost

One of the primary matters you will want to do while you make compost is to make certain that you have broken down any large count number into smaller pieces. Large chunks of kitchen waste, lawn waste and paper will take a very long term to interrupt down when you compost. It is constantly a good idea to shred paper and cardboard before adding and to ensure that whatever else you install is in small portions. Some human beings that compost choose to move ahead and shred all in their pieces earlier than adding in to shop time when they're geared up to feature to the bin.

Adding Ingredients All at Once

Next, while you are following a composting recipe, you'll need to ensure that you upload in all your ingredients at one time. This is due to the fact whilst you upload in elements a touch at a time, they'll not decompose at the identical price. It is not advisable to hold adding substances into your composting bin at exceptional instances. If you're involved approximately having waste sitting around watching for a composting bin, you then would possibly want to don't forget having several smaller composting areas going always.

Adding within the Right Order

As quickly as you are prepared to begin constructing your compost pile, you have to recognise that the order in that you upload your elements will determine your end results. You may have better compost while you upload it in layers—in a positive order. Start with an awesome thick layer of twigs and small sticks. This will boost the bottom of the compost up just barely and allow for circulation and a chunk of respiration room. You will then want to layer the carbon fabric after which the nitrogen material in alternating layers. It is continually a very good idea first of all the carbon, then nitrogen and so on till you've got all of it introduced into your compost bin.

Water Comes Next

Moisture, in the proper amounts, is important to composting. You can't have desirable compost with out water. Once you have got all of your layers in your compost bin, you will need to feature water. You will need to make certain that you do now not add too much. You will upload simply sufficient water to make the pile sense a chunk like a damp (but no longer soggy) sponge. Imagine the sensation of a sponge that has been squeezed out. That is the sensation you are going for whilst you compost. One certain-fire way to tell if you have enough water (or an excessive amount of water) to your compost pile is to get a small handful of the aggregate. Squeeze it on your hands. Does water drip out? If only a few drops of water comes out, you then have the precise quantity of water. If water pours out, then you definitely have brought an excessive amount of and you will want to feature extra dry components to the aggregate to saturate the excess water. If no water comes out in any respect, then it's far too dry.

Let Your Compost Bake

The amusing part of composting starts offevolved after you've got delivered in all your ingredients and the water. It is time to stir and wait. You will want to make the effort to stir and rotate your compost every 5 to seven days. This will make certain that each one of the elements you have layered into your compost pile or bin is calmly distributed and circulates well. Stirring is likewise vital for your quit result because it will replenish the meals and oxygen for the microorganisms that are in your compost. Those microorganisms need a sparkling supply of each every few days. Stirring will even assist ensure that the temperature of your compost is calmly distributed. It is vital that the temperature of the mixture remains among 104 stages to a hundred and sixty ranges. This should be the case even in colder climates. One interesting observe about temperature: when the temperature reaches around one hundred thirty ranges, maximum of the ailment and pests, together with seeds and weeds will die internal of your compost. A top temperature may even assist ensure that your composting will decompose faster.

Another be aware about composting: if you choose a low preservation composting recipe, it's going to take longer because the temperature will not reach a higher temperature than the ones which can be high maintenance. These will no longer, but, kill off seeds and weeds, so don't upload those to the low protection composting.

When is the compost executed?

Of direction, when you spend the time making compost, you'll also need to recognise while the compost is finished. After you've got added all your elements to your compost bin and stirred and permit it sit for the best amount of time, you want to know when the compost is prepared to apply. It is generally easy to inform while it's far "baked". Most of the time, your compost aggregate is finished whilst it has an earthy scent. All of the particles which might be inside the mix must appearance the same size and texture. In addition, compost that is prepared to apply may have the shade of darkish brown soil and could experience light and fluffy. There must not be a nasty or bizarre smell to compost while it is done and if you see that your composting fabric has now not broken down into uniform texture and length, then it desires to take a seat some time longer before being used. Making positive that your compost is prepared will come up with beneficial compost.

Testing the Compost

Another way to tell whether or now not your compost is prepared to apply is to see if any of the original materials nonetheless looks as if what it become whilst it went in. If so, your compost is not equipped to use yet. If you are nonetheless now not sure if your compost is prepared to take out and use there is any other simple take a look at that you could perform. Expert composters endorse disposing of a handful of the compost and putting it in a small plastic bag (a thick zip pinnacle freezer bag will paintings well). Close the bag tightly and permit it take a seat for 24-forty eight hours. When you open the bag, there need to be no strong scent. If no longer, this means you may start the usage of your compost. The excellent stop end result that you could wish for with composing ought to have a mild, nearly sweet earthy odor and have a pleasant mild texture.

It might also take a little work and time with the intention to learn when the composting is ready to use, however once you recognize the distinction among "appropriate" compost that is ready, and compost that needs extra time to "bake", you'll soon be an professional at making your compost.

Chapter 2: Making A Worm Composting Environment

To make your malicious program composter you first off need to discover a field you may use with a decent becoming lid – you don't need the ones wee beasties escaping, now do you?

For a typical two to 3 character family who are doing away with round eight kilos (3.6kg) of food scraps each week a metallic, plastic or maybe wooden bin this is two toes huge, four ft long and one foot high will do.

As a hard rule of thumb, the field you're the usage of needs to offer one rectangular foot for each pound of food scraps you'll installed it each week. The bin desires to have drainage holes at the bottom with a tray to catch the compost tea (recollect that?). Your bin additionally needs to have air vents on the edges and the pinnacle.

Worms thrive in a damp, dark environment and are made up of among seventy five% and 90% water. They want to be wet so as for them to breathe. You want to cowl your bug bin to make it damp and dark. Inside you can placed a few darkish plastic or cardboard on pinnacle of the bug bedding and outdoor you operate a stable lid if you want to maintain out pests and the climate.

You can keep your trojan horse bin anywhere you need, though it's far glaringly extra convenient if it's miles somewhere out of your way. It wishes to be saved out of the way of extremes of temperature and heavy rain. Worms do first-rate in temperatures within the 70's (Fahrenheit) though they'll be ok between 40F and 80F. If the temperature drops underneath 40F you then need to either pass your computer virus bin interior or ensure it's miles well insulated to preserve your worms warm.

Whilst you can make your very own computer virus bin from a shop sold container, you may should drill holes in it for drainage and air plus need to give you a method to catch the compost tea with a purpose to be produced (this makes an awesome liquid feed so don't dispose of it). For many human beings a commercially made computer virus bin is a lot easier and much less hassle.

If you make a decision to construct your worm composter outdoor then ensure it is nicely covered in any other case you may discover scavengers and rodents trying to interrupt in to devour your worms and kitchen scraps.

Once you've got created your bin you need to subsequent get the bedding together on your worms. This can be product of shredded newspaper (not smooth paper even though) and corrugated cardboard (again, not sleek). You can use shredded fallen leaves, seaweed, dried grass cuttings, aged manure, chopped up straw and so on. It relies upon on what is to be had for your local place as to which you'll use.

There are commercially to be had worm beddings to be had which you may buy online or from a lawn save near you.

The bedding wishes to be moistened. The easiest way to do that is to put the bedding right into a big field and then cover it with water. Leave it to soak for a few hours so that it is able to soak up the water.

You may also must then squeeze out the bedding. It wishes to have a consistency of a properly-wrung sponge in place of be dripping moist.

Once the bedding is dampened you need to fill your worm bin approximately two thirds full with it. A two foot by way of 3 foot bin will need among ten and fifteen kilos of bedding.

When you positioned the bedding into the bin you do not want to push it down. You want to depart some area between the bedding for air, to be able to allow the worms to move round, preserve the aggregate aerated and forestall your bug bin from smelling bad.

You will want to frequently test your worm bin and make certain that it's far nonetheless damp enough. If it is drying out then damp it down once more using a twig bottle in order that the computer virus bedding is damp, like a wrung out sponge once more.

It is really worth scattering multiple handfuls of soil and sand within the bedding. This provides grit, which allows the trojan horse's to digest the food you are putting in their bug bin.

Now your worm bin is subsequently ready for some worms. You need to make certain you get the right sort of worms – they must be redworms due to the fact these are the first-class at breaking down organic depend. Do now not use the worms from your garden on your bug bin due to the fact they may be not suitable and could now not ruin down the kitchen scraps well.

You can get redworms from lawn stores or on-line. You can also often find them in fishing bait shops too. If you need to get them without cost and don't thoughts getting your hands dirty then you may be capable of discover them in any well-rotted manure.

The subsequent query you've got is how many worms you'll need. As a rough guide you want among one and two kilos of worms for each pound of meals waste your family produces each day.

If you start out with too few worms, don't worry, they will multiple very rapidly and you will quickly have more! Just alter the amount of meals waste you are installing there till their variety will increase.

Your worms are speedy breeders and they can double their population each ninety days! Bear this in thoughts while you are designing your bin and filling it with meals scraps.

Just scatter the worms at the pinnacle of the bin and they may wriggle their way down into the bedding to get out of the mild. Redworms do now not like mild and could quickly burrow down in to the dark.

There are multiple procedures to including food waste in your trojan horse bin. You can just scatter it at the pinnacle, that is the very best way of doing it. The worms will eat the scraps and pull them down in to their bedding.

For those who decide upon greater order to their lifestyles you can mentally divide your bin in to some of exceptional sections and then, over a few weeks, bury food scraps in every segment. Just pull back some of the bedding after which dump the scraps into the hole you made. Then cowl those scraps with a few bedding.

The gain of this method is that if you paintings your way throughout the bin including food step by step, by the point you get back to the first location the worms can have quite an awful lot finished the composting process there.

If you be aware that your bug bin is starting to odor then you may want to either put in less meals or cut it up into smaller pieces. Generally worms prefer veggies to fruit and they're not so eager on citrus. However, supply them time and they may destroy the ones down too.

It will take twelve to sixteen weeks for the worms to absolutely digest the meals scraps and make the compost, additionally known as computer virus castings. These have 5 times greater nitrogen than your lawn soil, seven instances greater phosphorus and 11 times more potassium, that means they're very high in essential nutrients.

Remember the compost tea we mentioned in advance on? This is an terrific fertilizer that you can use to your indoor and outside plants. Make positive which you have a way of extracting this from the bug bin due to the fact if it leaks directly to the floor it'll depart a stain this is very difficult to take away.

The worms are very touchy to mild so if you shine a brilliant light into the bin they may all flee in to the lowest of the bin to escape from it. You can then use your hands to dispose of the pinnacle layer of the malicious program casting.

You can then start to dispose of the bedding little by little and the worms will hold to dig deeper and disguise. As you are casting off the bedding you want to pick out out any worms which can be in it and go back them to the bin. You may find some computer virus eggs, which might be small cocoons which are opaque. These want to be lower back to the malicious program bin. Once you are accomplished you can refill the bug bin with layers of moist bedding and begin the procedure again.

Another method is to push the bug castings to at least one aspect of the bin after which fill the alternative aspect with new bedding and kitchen scraps. Leave it for some days and the worms can have vacated the vintage bedding and be taking part in the brand new bedding and food you've got given them. Then all you do is do away with the finished compost and pick out worms and eggs as earlier than.

If you prefer you could tip your bug bin out on to some plastic sheeting and kind thru the pile. Move the malicious program castings to one side and placed the worms and their eggs back into the bin with some new bedding and food scraps.

It is quite clean to manipulate a trojan horse bin and it's far a great way with the intention to generate compost in a constrained space. It is compact, neat and smell unfastened and, aside from the emptying of the bin, pretty a great deal hard work loose too. Worm composting produces high quality compost that is very precious to absolutely everyone who grows flora indoors or outside.

Chapter 3: Composting Basics

Five principle territories that ought to be controlled for the duration of composting are:

Feedstock and Nutrient Balance

Composting requires the ideal equilibrium of green natural materials and earthy coloured (brown) organic substances. Green natural fabric includes food scraps, grass clippings, and manure, which contain loads of nitrogen. Brown organic substances comprise wooden chips, branches and dry leaves, which include a whole lot of carbon but little nitrogen.

Getting the right nutrient blend requires experimentation and patience.

Temperature

Microorganisms require a specific temperature range for ideal movement. Certain temperatures strengthen short composting and annihilate pathogens and weed seeds. Microbial action can raise the temperature of the heap's center to at least one hundred forty°F. On the off risk that the temperature does no longer boom, anaerobic situations (i.E., rotting) appear. Controlling the past four elements can obtain the ideal temperature.

Moisture Content

Microorganisms dwelling in a compost heap need sufficient moisture to undergo. Water is the key element that facilitates transports materials within the compost heap and makes the nutrient in natural fabric open to the organisms. Natural fabric incorporates some moisture in converting sums, but moisture additionally may come as rainfall or planned watering.

Particle Size

Granulating, chipping, and shredding substances builds the floor territory on which microorganisms can feed. Smaller particles additionally produce a greater homogeneous compost mixture and enhance heap insulation to assist keep up best temperatures. On the off danger that the debris are excessively little, though, they may hold air from streaming brazenly thru the heap.

Oxygen Flow

Putting the heap on a chain of pipes, turning the heap or which includes bulking agents for example, shredded newspaper and wooden chips all help circulate air through the heap. Circulating air via the heap lets in decay to show up at a faster price than anaerobic conditions. Care ought to be taken, nevertheless, now not to give loads oxygen, that could dry out the heap and preclude the composting procedure.

Onsite Composting

A institution of human beings or a own family which are going to compost small quantities of wasted food can compost onsite. Composting can altogether reduce the amount of wasted food this is discarded. Little amounts of food scraps and yard trimmings can be compost onsite. Enormous amounts of meals scraps and animal merchandise are not appropriate for onsite composting.

Onsite Composting

What you want to realize approximately Onsite Composting

Food scraps must be taken care of correctly in order that they do no longer purpose odors or pull in unwanted animals or bugs.

Onsite composting takes subsequent to no time. Training is the important thing and it calls for much less gadget. Local groups may additionally keep composting shows and seminars to induce homeowners or organizations to compost on their very own houses.

You can set leaves to the facet and use them as mulch round timber and scrubs to hold moisture.

The environment and seasons adjustments will now not bigly have an effect on onsite composting. Little modifications can be made while modifications take place, for instance, when the wet season attracts near.

Making compost can require as long as two years, however guide turning can accelerate the process to between 3 to a half of yr.

Compost, be that as it can, ought not be applied as gardening soil for houseplants because of the presence of weed and grass seeds.

You can depart grass trimmings on the lawn-known as grasscycling. These cuttings will crumble normally and return a few nutrients lower back to the soil, like composting.

Vermicomposting

Red worms in containers feed on backyard clippings, meals scraps and other natural count number make a distinction to make compost. The worms decompose this material into excellent compost referred to as castings. Worm canisters are not difficult to assemble and are additionally on hand for procurement. A pound of mature worms (kind of 800-1,000 worms) can consume as much as a massive part of a pound of natural cloth every day. The bins may be sized to match the quantity of food scraps with a purpose to be transformed into castings. The castings can be utilized as potting soil. It often takes three to 4 months to deliver usable castings. The other result of vermicomposting referred to as bug tea is utilized as a excessive nice liquid fertilizer for gardens and houseplants.

Vemicomposting

What Can Be Composted - Vermiculture?

Yard clippings like flowers and grass

Food scraps

Paper

What you want to recognize approximately Vemicomposting

The fine temperature range for vermicomposting is from fifty five°F to seventy seven°F.

It is vital to maintain the worms alive and healthy via imparting ok meals and appropriate situations.

Perfect for condo dwellers.

Vermiculture can be use by schools to train children recycling and conservation.

In hot, dry regions, the box should be set underneath the colour.

Uttermost temperatures and direct daytime are not beneficial for the worms.

Prepare bedding, cover trash, and preserve away worms from their castings.

Vermicomposting inner can live away from a huge wide variety of those troubles.

Worms are sensitive to adjustments in surroundings.

Turned (Aerated) Windrow Composting

Turned windrow composting is suitable for large volumes, as an example, that created via whole network and amassed via neighborhood governments, and excessive quantity meals-dealing with organizations (along with cafeterias, eating places). It will yield big amount of compost, which may additionally require assist to marketplace the very last product. The government may additionally want to make the compost accessible to inhabitants for a low or no fee.

This type of composting consists of shaping natural waste into rows of lengthy thousands known as windrows and circulating air through them now and again via either automatically turning the lots bodily or manually. The best heap height is someplace in the variety of four and 8 ft with a width of 14 to sixteen ft. This length heap is sufficiently good sized to supply enough heat and maintain temperatures. It is sufficiently small to allow oxygen circulate to the windrow's middle. Enormous volumes of varied wastes like animal byproducts, (as an example, rooster and fish wastes), oil, beverages and backyard clippings can be composted through this technique.

What you want to know approximately Turned Windrow Composting

A liquid referred to as leachate is discharged at some stage in the composting system. This can sully floor-water materials and nearby ground water. It should be gathered and handled.

Windrow composting can work in cold environments. Regularly the outside of the heap may also freeze, but in its middle, a windrow can arrive at a hundred and forty°F.

Odors moreover ought to be managed. The public must be educated regarding the pastime and have a manner to cope with any grievances about odors and animals.

Windrow composting is an full-size scope activity and may be problem to regulatory enforcement, siting requirements. Compost have to be tested in a research middle for bacterial and heavy metal content.

During wet seasons, the states of the heap may be modified in order that water runs off the top of the heap as opposed to being ingested into the heap.

Windrow composting usually requires considerable plots of land, robust system, a good deal labor to function and preserve the power, and staying power to explore exclusive avenues concerning unique substances combos and turning frequencies.

Windrows are from time to time covered or set beneath a secure residence to maintain water from evaporating in a heat and arid weather.

Aerated Windrow Composting

Aerated Static Pile Composting

This type of composting produces compost reasonably unexpectedly (interior 3 to a 1/2 yr). It is suitable for a typically homogenous combination of organic waste and paintings perfectly for large quantity generators of backyard clippings and compostable municipal strong waste (e.G., meals scraps), like farms, gardeners or nearby governments. This approach, notwithstanding, does now not paintings well for composting byproducts of animals, oil or grease from food processing industries.

Organic wastes are jumbled together a huge heap in an aerated static pile composting. To aerate the heap, layers of loosely heaped bulking retailers consisting of shredded newspaper, wooden chips are brought in order that air can pass from the base to the top of the heap. The hundreds likewise may be positioned over a community of pipes that carry air into or draw air out of the heap. Air blowers can be activated by way of a temperature sensor or a timer.

What you want to recognise approximately Aerated Static Pile Composting

This technique may want first rate expense and technical assistance to buy, deploy, and maintain device like pipes, fans, blowers and sensors.

Applying a thick layer of finished compost over the heap may additionally help lessen any odors. In the occasion that the air blower draws air out of the heap, filtering the air through a biofilter produced the usage of prepared-made compost will likewise lessen any of the smells.

As there is no actual turning, this technique requires careful staring at to assure that the out of doors of the heap heats up as a whole lot as the middle.

Having a managed stockpile of air permits creation of large thousands, which require less land than the windrow method.

In a heat, dry surroundings, it is able to be important to cover the heap or spot it below a shelter to keep water from dissipating.

In the cold, the center of the heap will retain its warm temperature. Air circulate can be more difficult in light of the truth that passive air flowing is utilized in preference to energetic turning. Putting the aerated static lots inside with proper air flow is additionally in some cases a choice.

Aerated Static Pile Composting

In-Vessel Composting

In-vessel composting can manage quite a few waste with out occupying but lots space that the windrow method and it may accommodate practically any sort of natural waste (e.G., meals scraps, animal manure, meat, biosolids). This approach involves feeding of organic materials right into a concrete-covered trench, drum or silo. This permits outstanding manage of the environmental situations like airflow, moisture and temperature. The material is mixed or robotically became to make certain it's miles aerated. The length of the vessel can range in capability and size.

This method produces compost in only a few weeks. It requires a couple of extra weeks or months until it is prepared to utilize considering the microbial motion needs to stability and the heap wishes to chill.

What you want to recognise approximately In-Vessel Composting

Uses substantially less land and bodily paintings than windrow composting.

Very little scent or leachate is created.

Utilize in very cold weather is manageable with indoor use or insulation.

This technique is steeply-priced and can require specialized mastery to work it as it should be.

Some in-vessel are extremely significant, like the size of faculty bus. Big meals processing plants usually utilize these.

Some are adequately little to suit in a college or eatery kitchen.

Careful control, often electronically, of the surroundings lets in all yr utilization of this method.

In-Vessel Composting

Chapter 4: How To Begin Composting

Composting consists of blending yard and circle of relatives unit herbal waste in a heap or receptacle and giving situations that support disintegration. The deterioration procedure/gadget is energized by way of a wonderful many tiny organic entities (microorganisms, organisms) that take up residing arrangement interior your manure heap, continually consuming up and reusing it to supply a wealthy natural manure and giant soil. Is it Sound entangled? It's genuinely now not. Everything you need to reflect onconsideration on composting is an essential comprehension of multiple fundamental standards, and a tad bit of elbow grease. Nature does the relaxation. Decomposition, or the composting technique, takes place constantly and step by step round us each day. The somber, wealthy soil masking the forest floor is a extremely good pattern of this. The moment we compost, all we're absolutely doing is accelerating Mother Nature.

Area and Appearance: To begin with you'll must select your region for composting. Where you placed it is predicated on upon capability and aesthetics.

Regarding appearances and awesome family members along with your acquaintances, you most probable would pick now not to place your compost box to your the front grass alongside the letter box

Rather, pick out the terrace, or, on the off chance that you don't have one, then a receptacle located for your cellar can do the trick.

Need to assemble your personal? Here's one fundamental association: exchange over old delivery pallets (which you could generally get free of rate) into a compost "storehouse." Use one for the base. Pound in metallic bolster shafts and later on consist of beds through slipping them over the bolster posts to make your canister's dividers and you are proper to go. The University of Missouri Extension offers some specific instances for building a canister.

You can likewise pass the receptacle (a shape isn't essential) and in reality have a compost heap or stack. In regards to appearances — and if your homeowners affiliation is fastidious — you may need to screen the heap from perspective by using planting timber or a wall. You'll moreover presumably now not need it by way of your park or distinct degrees outdoor wherein you engross.

From a practical perspective, you will require a niche with first rate air flow. Don't positioned it along your property or other wooden structures as the deteriorating scraps and coming approximately compost may additionally cause the wood to decay. Incomplete color is a smart idea so the compost would not get overheated. Additionally affirm the spot of location in which you put your stack gets terrific seepage.

Near to the backyard and to a water source are each terrific spots for building a compost bin on account that it will be simpler to transport the materials to and from the enclosure and much less worrying to water it. An exchange thought may be to vicinity it close to your kitchen to make it beneficial to region desk scraps on the heap or inside the container.

Size: Make certain your pile isn't always littler than 3' x three' x 3'. Actually, that is likely the proper length. It's sufficiently correct sufficient to "cook dinner" your waste and alternate it into manure, and now not all that massive that it will get to be unmanageable and hard to show.

Dampness: The microorganisms that do your grimy work within the compost heap wishes water for survival, but it may be hard to choose how much water to encompass and whilst. An immoderate amount of water means your natural waste won't collapse and you'll get a foul and rank heap that could properly response to the call "marsh thing." Too little water and you will execute the microorganisms and you will not have your compost.

One preferred guiding principle: the greater green cloth (grass, weeds e.Tc.) you put internal you compost bin, the much less water you will should include. Actually, at the off hazard that you need to encompass dry items, for instance, straw or roughage, douse the cloth first in water so it will not dry out your compost heap. Usually your compost should be clammy and not sopping wet.

On the off threat that you are composting at home and you get a ton of downpour, fabricate a top over the heap. This may be as smooth as a protecting. The purpose you need to offer your compost heap extra safe residence is as a consequence of vitamins, or leachates, linking out in the course of downpour. That isn't always such an trouble in a niche where precipitation is not substantial, however inside the occasion which you get a full-size measure of downpour in which you stay, it can have a chief impact. An extra of water inside the heap will ease off the technique and might likewise make it foul and disgusting.

Air flow: Oxygen is also wished with the aid of some of the microorganisms in price of powerful composting. Provide them sufficient ventilation and they'll deal with the rest. You can verify that the microbes to your compost get sufficient air via turning the heap frequently. Utilize a pitch fork, spade or manure aerator to mixture your heap. In the occasion which you've got a manure tumbler, you have it simple. Simply wrench that lever. Don't flow into air through your manure and it's going to separate progressively, bringing about a disgusting, thick, smelly heap. It's likewise a clever notion to turn the substance since it reworks the rotting material. With a touch care, you could flow the much less deteriorated fabric on the edges to the center of the heap to warmness up.

Turning your manure with a simple to-make use of manure aerator is a viable method to include oxygen and bring microorganisms into contact with these days protected material. Basically push the slicing stop into the heap. As the apparatus is withdrawn the pivoted oars open out, circulating air through and mixing the heap — without tough paintings!

Temperature: As they devour, the living organisms in rate of composting create numerous warm temperature, which boost the temperature of the heap or manure canister and speeds up deterioration. A compost heap this is functioning admirably will create temperatures of 140-160°F. At these temperatures all weed seeds and plant sicknesses can be removed. An "tremendously warm" manure heap will produce temperatures of as much as 170°F for as much as every week or more. Utilize a manure thermometer to gauge the accurate temperature at various regions inside the heap. The composting thermometer indicated here consists of three temperature zones on the dial that will help you supply the finest "black gold" in your backyard. A simple comprehension of the temperature in a heap will assist you apprehend:

When its completed

 When to consist of extra substances

When to consist of water

When to turn it

Adding compost Materials: At the factor while you are adding waste in your compost, do no longer weigh down the compost items down a good way to make greater area to include more. Squashing the substance will press out the air the ones microorganisms in the manure heap need to transform your waste into gold. (Rather you may be encouraging the anaerobic organisms, which moreover do a awesome activity converting over carrot peels and other natural be counted into manure but have a propensity to be smellier.)

Likewise be prudent about filling your receptacle. Incorporate a combination of chestnut stringy fixings and greens. A decently adjusted "eating habitual" will guarantee that composting does not take too lengthy and that you don't wind up with a disgusting, rotten stack. Additionally shred, shakers or generally make scraps littler, so as to assist the inhabitant microorganisms, do a outstanding process in changing over the junk into compost.

At final, after you have protected kitchen vegetable waste, toss a few leaves or grass clippings on pinnacle of it. This will assist maintain things adjusted, lower smells and make your fertilizer canister much less attractive to critters who are attempting to sniff out a free dinner. As organic objects in a manure heap warms up it splits down and consumes up less room. A manure heap can decrease as much as 70%.

Chapter 5: Decomposers – Their Role In Composting

Over time, a compost pile will become a veritable haven for flora and fauna and a few human beings even locate slow worms buried in the decomposing plant remember. There are many distinct sorts of lifestyles lively in your compost heap and it's far essential that you apprehend them so that you do not worry when you see insects creeping around on your compost. The only motive you don't see bugs in keep bought compost is because the compost is generally sterilised to eliminate the bugs to make it more appropriate to shoppers.

Firstly, there are chemical decomposers including fungi, micro organism and actinomycetes, which are gram tremendous, anaerobic micro organism that grows into branches with filaments, growing an in depth colony or mycelium. Most of the decomposition in any compost pile is due to those organisms.

Then you have large, bodily decomposers inclusive of mites, centipedes, millipedes, snails, spiders, slugs, beetles, ants, flies, worms, timber lice (tablet bugs) and extra. These eat the decomposing plant material and spoil it down into smaller elements. Even the a lot hated slug and snail have a purpose inside the decomposition process.

Without both the micro-organisms and the larger decomposers, the composting method simply wouldn't paintings. Spores and micro organism are all round us and, given the proper conditions, grow very quickly. Depending on the temperature of the compost pile and what's in it, one-of-a-kind micro organism and insects will take in residence inside the compost. Over time, as conditions alternate, those develop, die and develop once more.

In a compost pile, there will also be fungi developing. Much of this can no longer be visible to the naked eye, whereas occasionally you'll see mushrooms seem on top of your compost. Don't fear if this takes place as it is flawlessly herbal and a part of the composting method. However, do not eat the mushrooms as many species are toxic and even lethal. A lot of humans worry that the mushrooms imply some thing is inaccurate while they actually suggest the composting process is progressing properly.

Fungi growing on composting manure

Eventually, worms make their way into the compost and be a part of within the decomposing method. If your compost pile is not immediately at the soil, the worms can also war to get into your compost so you can upload some whilst you find them inside the lawn. If your compost is direct on the soil, the worms will evidently come up from the soil and into the compost pile. Again, don't worry in case you see them as they are beneficial and disappear into the soil whilst you operate the compost.

Now you recognize a bit more approximately how a compost pile becomes a micro-organism zoo through the years. Don't worry in case you see bugs or fungi, just hold turning your compost and it is going to be truely quality.

The Different Types of Compost Bin

There are masses of different sorts of compost bin, and this chapter explores the numerous options for home composting. Your selection depends to your price range, abilties, and the to be had area. Ideally, you want a compost pile this is big sufficient to comprise your waste for 4 to six months and with enough spare space to turn your compost into. In exercise, you are more likely to be limited in area and want to select something this is suitable in your lawn.

Plastic Standalone Bins

These are the maximum normally discovered compost containers in the lawn and are for continuous composting, i.E., you upload new fabric to the top even as those at the lowest keep to break down. Some devices are just enclosed plastic containers, while others have air vents on the perimeters. Some are closed at the lowest while others are open to the soil. Most have some kind of vent or door at the bottom to benefit get entry to to the completed compost.

When deciding on this kind of compost bin, you need a decent-fitting lid, correct and clean get admission to to the completed compost and one that is at least 3 toes in diameter. The lids can blow off in high winds, so weigh them down with a brick or something comparable. When empty, the bins are very mild and also can blow away. This form of bin may be dug a couple of inches into the soil to cause them to steady in high winds.

As these are messy and awkward to transport when full (accept as true with me, I've executed it), they may be some distance better to be sited of their final position before you start filling It with plant cloth. Make certain they are on level ground with out a gaps among the bottom of the bin and the floor as that permits rodents to get into your compost.

It is a good deal harder to show compost in this form of bin, but there are expert tools available. Unless you buy the sort of, you're poking a stick or within the bin and wiggling it around to aerate the compost. Compost takes longer to shape on this type of bin due to the fact you cannot without difficulty flip it. For the home-gardener, this is right as it's miles neat, can be placed unobtrusively and could be big sufficient for the small to medium sized lawn.

Tumbling Compost Bins

These are excellent for making a batch of compost instead of for continuous composting. Basically, you gather your compost and then throw it all inside the tumbling bin at the same time. That way while your compost is prepared, it's going to now not have uncomposted be counted in it.

These are without a doubt appropriate, even though they often make a good addition to a multi-bin compost pile as those can produce compost in much less than five weeks with a very good quantity of nitrogen cloth and masses of turning. With it only taking a few seconds to show, it is very clean to apply and a wonderful way to regularly produce fresh compost throughout the year.

A compost tumbler is desirable to the home-gardener or as an addition to composting on a larger scale. It can't be sited as unobtrusively because the previous sort of bin as you want get right of entry to to turn it, but it is able to nonetheless be positioned out of the manner someplace.

An example of a tumbler that may be offered on line

Wire Bins

You can without difficulty make a compost bin with a few cheap materials from your local hardware or DIY store. Buy an eleven foot duration of 2" x 4" x 36" welded, medium gauge fence twine and tie the ends collectively to shape a cylinder. Push this into the floor, and you have a bin that could keep round a cubic backyard/metre of material!

You may want to make or 3 of these bins collectively, after which fill two containers, leaving the 0.33 unfastened to show the compost into. The handiest problem with these can be leaning in to show or get entry to the compost. However, if you can unhook the twine fence so it turns into flat again, then it's miles simpler to get to the compost.

Trash Can Bin

Another easy compost bin to make via slicing the lowest off a garbage bin and drilling holes around the outside for aeration. Bury the lowest of the bin some inches into the floor and press the soil along the edges to stable it in area and prevent rodents moving into. This enables useful micro-organisms from the soil get into your compost and paintings their magic in addition to maintain the bin solid.

Again, the drawback of this kind of bin is get admission to, however as these containers tend to be three to 4 ft excessive, it isn't too bad. With this form of bin, you don't generally turn the compost because of access, and also you normally fill it abruptly.

Brick, Block or Stone Bin

You can build a compost bin from vintage bricks; you can buy them or frequently get them without cost from a building website online or residence preservation. Lay the bricks to shape three sides of a rectangle round four toes rectangular, leaving one aspect open for ease of get right of entry to. Make the partitions three or 4 toes high and both without or with mortar. Make sure you leave normal areas in the wall for aeration. Leave half of brick gaps and use mortar to make sure the shape is sound and gained't fall over while turning the compost. You'll be surprised how smooth it's far to knock an unmortared wall with your fork or lean on it and for it to fall down.

This is a fantastic kind of compost bin to make and can be made the use of loose or recycled materials. You can make a gate for the the front out of a wood pallet or other material to preserve the compost in region as you fill it. If you may get pre-shaped concrete fence panels, those are appropriate for this form of compost bin too.

A compost bin built from antique doorways

This sort of bin is appropriate for the bigger home gardener, the allotment proprietor and even the smallholder. Build a couple of packing containers like this and you have lots of space on the way to suit the bigger land holder.

Wooden Pallet Bin

This compost bin is simple to make from recycled or loose substances. Many factories and shops will happily come up with pallets without spending a dime as it saves them doing away with them. I've discovered them on commercial estates earlier than and agencies were glad for me to take them away in place of them have to pay for them to be disposed of.

You virtually shape a cube with pallets, leaving the pinnacle and the the front open. Putting a pallet on the bottom facilitates with airflow, although it stops worms coming up from the floor into the compost. Line the inner of the pallet with heavy, black plastic with some holes in for aeration to maintain the warmth in and stop the compost falling out of the gaps between the slats within the pallet.

These are first rate and, if you have the gap, then I propose constructing a or 3 pay compost gadget. Use an additional pallet as a gate at the the front; I'll show you the way to make one of these a bit later on this e-book.

This sort of bin is famous with allotment holders, but less common at domestic because of aesthetics and space requirements. However, if the outside of the pallets is painted and the packing containers positioned faraway from any seating regions, these are excellent for a gardener with a larger quantity of fabric to compost.

Multi-Bay Bins

These are the high-quality type to have and can be made from bricks, pallets, timber or cord, relying at the sources you have to hand. In a three bay set up, you'll generally have in use, and the 0.33 is where your grew to become compost is going. It makes turning compost a whole lot simpler, and you can be as creative as you need to construct gates, hinged lids and so on if you want.

On a larger scale, which includes a smallholding, multi-bay bins may be constructed from dealt with fence panels to make huge enough bays. Using six or eight foot fence panels manner it is feasible to get a small digger into the bays without problems enough to turn or move the compost. When composting on a scale like this, it may be well worth constructing a roof over your compost so that you can manage the quantity of water that receives to it and to keep the location better enclosed.

When it comes to making a compost bin, you could use your imagination and creativity. They can be as simple or as complex as you want, so long as they permit clean access to turn and remove the compost plus have sufficient room for everything you need to compost at some stage in the yr. Look on the vicinity of land you have and have a bet at how lots compostable fabric you'll produce, and then double it as you will usually discover more matters to compost.

The Composting Process Explained

The waste materials you throw on for your compost pile are was nutritious compost via fungi, enzymes, micro organism, bugs and other micro-organisms, most of which can be too small for you to see. Your job is to make certain that they have got the nice feasible environment in which to perform their feature in existence and bring your compost. Even if you don't offer the perfect surroundings, you continue to get compost, however it is able to take a yr or extra to appear, as compared to the numerous months if the situations are ideal. Generally, humans don't have time for 'best' so pass for as precise as you could in the time available to you.

To supply the composting procedure the great situations you want to make certain you have sufficient of those 4 things:

Carbon – carbon rich substances provide strength for the micro-organisms on your compost heap. These are known as brown materials and consist of factors like straw, dry leaves and shredded paper, as discussed formerly.

Nitrogen – that is the meals that shall we the micro-organisms grow and multiply swiftly. Weeds, grass clipping, kitchen scraps and different inexperienced rely are all complete of nitrogen. You can upload manure, seaweed and blood and bone meal if you need to present your compost pile greater nitrogen.

Water – too much water will drown the micro-organisms, but too little will dehydrate and kill them. Ideally, your compost wishes to be as damp as a properly wrung sponge. To upload water to a compost pile, put a lawn hose into the middle of the pile in several locations and squirt in some water. Alternatively, flip your compost pile and sprinkle it with water. Un-chlorinated water is taken into consideration the nice for a compost pile, but it isn't some thing most human beings need to hand. Make your personal, by using leaving a bucket or two of water sitting in a single day for the chlorine to evaporate off. Normal water will do just exceptional in case you haven't were given time to leave buckets of water status round. Covering your compost pile enables prevent lack of moisture and hold heat, which speeds the composting manner up.

Oxygen – the magic is finished via residing organisms which, such as you and I, want oxygen to live to tell the tale. When you first build your compost pile, there may be masses of gaps among the substances containing oxygen, however because the plant be counted breaks down the oxygen is used and those gaps disintegrate as the plant material rots. Unless you switch your heap and introduce oxygen, the micro-organisms will in the end suffocate and the decaying manner will slow down till it sooner or later stops. Turning or aerating your compost pile gets oxygen back into the pile so it can hold to interrupt down. If you have got left your compost a while and it has long past slow like this, then when you switch the pile, dig in a shovel or of soil or manure to re-introduce beneficial organisms into your compost. This kick starts the composting method and get the plant matter breaking down again.

When Your Compost Is Ready to Use

Compost is ready to use when it turns into crumbly, darkish brown and has a totally earthy odour about it. It ought to now not be rotten or mouldy but can be fluffy in preference to powdery, which means it is 'overdone.' You received't be capable of see any of the authentic substances in your compost except some large woody portions. It have to be at the ambient temperature of the vicinity it is in and must not reheat whilst became.

Only a gardener could get excited with the aid of sparkling compost!

You may additionally notice worms and different insects to your compost now the temperature has dropped to a stage in which they are able to survive. If your compost smells of ammonia, is still warm or you may see a great deal of the original material then it isn't geared up to use.

When your compost looks as if it's far finished, dig it out and depart it for two to 3 weeks to ensure it has finished composting and allow the bugs to vacate the completed compost. Don't be tempted to apply the compost earlier than it is ready as it can grow to be hurting your vegetation as the composting micro organism will compete with the flora for nitrogen in the soil causing the plants to be stunted and yellow. A 'warm' compost should become burning delicate seedlings and prevent seeds from germinating.

The composting method is very simple and know-how the 4 important components of a compost heap guarantees the waste substances smash down quickly into useful compost. However, don't get too caught up in it all, as in fact, it isn't always generally feasible to get the balance ideal. Keep a bale of straw by means of your compost pile and upload a few every time you upload inexperienced fabric to assist the composting manner.

Some lovable finished compost made in a one ton bag

Chapter 6: Vermicomposting

Vermicomposting, or bug composting, is a form of composting using purple wiggler worms because the number one decaying agent. The worms' waste materials or castings are taken into consideration as high value fertilizers.

One advantage of vermicomposting is that the crimson wiggler worms aren't truly very picky with regards to food. They can consume meals scraps or even shredded paper. As a rule, the worms can consume anything remotely fit to be eaten.

Worm boxes are also particularly smooth to construct. In truth, many of these are without difficulty to be had in home improvement depots. These bins are compact, which makes this sort of composting ideal for small offices or flats.

Furthermore, the malicious program castings make extraordinary potting soil. This soil is commonly loose from artificial substances, dangerous chemicals, plant or animal illnesses, soil pests, and elaborate weeds. It additionally has the very best awareness of natural nutrients.

Worm tea is a liquid form of soil conditioner this is produced by means of worm compost. This is normally most beneficial to house or indoor plant life.

Unfortunately, the crimson wiggler worms want a consistent living temperature that allows you to thrive. They do now not fare well in regions that constantly enjoy sudden shifts inside the climate. These creatures need everyday feeding and frequent temperature test, in particular if the trojan horse containers are located exterior. The worms need to be protected from possibly predators as well, inclusive of birds, moles, or rats.

Indoors, the malicious program boxes might want proper ventilation and cooling to prevent unexpected changes in temperature. The moisture stage of the bug beddings need to also be cautiously monitored to keep away from that musky, damp scent that could turn out to be overpowering in enclosed spaces.

If you are keeping the pink wiggler worms outdoors, those can come to be pretty sensitive to surprising temperature adjustments. High warmth from a totally hot day or a sudden drop in temperature can kill off the purple wiggler worms en masse, as with too much moisture from rain or fog. The most suitable temperature for wholesome pink wiggler worms is around 55°F to 77°F (13°C to 25°C).

Indoor bug preserving is a better alternative. But the worm bins should be properly ventilated, properly stored with barely moist shredded paper (the worms' preference of bedding), and tended with simply the proper amount of meals. Placing an excessive amount of food Inside the worm containers also can sell bacterial and fungal boom that would poison juvenile worms.

Likewise, the bins have to also have just the proper quantity of moisture to make sure that the worms remain hydrated. These creatures get their drinks from the compost materials.

This sort of composting most effective requires a terrific quantity of mature crimson wiggler worms, worm bedding like shredded cardboards or newspapers, the real worm packing containers or boxes, and any equipment that could separate the worms from their castings (i.E. Net or mesh).

One pound of mature crimson wiggler worms (approximately 800 to one,000 worms) can yield triple their weight in castings in 1 week. This means that this sort of composting can yield a lot of excessive fee materials in as low as three months. These castings are often used as potting soil.

How to Start Your First Vermicompost

One of the first-class methods of harvesting high high-quality fertilizers and soil conditioners is by using the usage of vermicomposting. Red wiggler worms are used because the number one agent of deterioration. These critters can without difficulty be taken care of in suitable containers at home. These worms without problems method maximum sorts of kitchen scraps and even paper-based merchandise.

Step via Step Strategy

1.

Try to discover a supplier of worms that you can use.

Reliable worm suppliers regularly sell their "merchandise" by their genus names, along with:

EiseniaAndreii, or more commonly called tiger worms.

EiseniaFetida. This is the most extensively to be had "composters" which can be additionally regarded with the aid of the names brandling worms, manure worms, reds, red wigglers, red worms, or wigglers.

LumbricusRebellus, or more generally called the European crimson malicious program.

It is important to use the worms which are native for your location/united states of america. This would prevent doubtlessly invasive species from multiplying and frightening the nearby environs, in case some of your worms get away. Also, local worms would acclimatize more quick to the neighborhood climate, and therefore thrive higher ultimately.

If you aren't specifically sure which genus of worms are native to your region, definitely ask other worm composting fans on your area what sort of worms they use. Or, you could additionally ask the suppliers which worms are offered the maximum for your town, town, or country.

As a home composting choice, you would best want 1 pound of worms. That would be around 800 to at least one,000 live crawlies. It is vital no longer to move overboard whilst buying worms. Otherwise, as a substitute saving money through composting kitchen scraps, leftovers, and excess meals, you might actually locate yourself spending extra money simply to feed your worms.

When you buy your composters, usually buy stay purple wiggler worms, and now not bug eggs, and truly not live bait worms.

2. Find a appropriate composting bin.

Worm composting boxes are effortlessly to be had in most domestic improvement depots. You can either use plastic or wooden boxes, but the latter is most desirable for first time users.

Wooden packing containers allow air to flow into more freely inside. This makes it simpler to control temperature and moisture ranges.

The size of your composting bin relies upon on how a lot meals or kitchen scraps you usually produce in a day. If you generally get by using with much less than 2 cups of kitchen scraps in step with day, you will best need a small composting bin. If you want to compost greater food objects, or have a larger extent of paper products to system, then purchase the medium-sized or big-sized composting bins.

You also have the choice of purchasing or more smaller boxes, in case you have very limited area, however a whole lot of raw substances to manner. Worm containers are often stacked in piles of twos and threes to store area.

Again, you don't want to shop for greater than 1 pound of worms. If you do have a number of natural substances to compost, these worms can take care of the "load" as long as they've enough space and soil to transport around in.

Worm composting bins are ideally shallow, because the worms thrive better within the topmost phase of the soil. So strive now not to buy the deeper containers. Also make sure that the packing containers have enough respiratory holes at the edges of the edges. These packing containers usually price among $50 and $150 according to piece. Some of the pricier ones can effortlessly value $three hundred or greater.

three. Create the trojan horse beddings.

Red worms choose damp, warm, however airy dwelling quarters. To create the beddings, you will need to:

Wash out the bins thoroughly with detergent soap and water. Make certain which you wash away all cleaning soap residue afterwards. Dry those out slightly before the use of, in particular around the respiratory holes.

Take a handful of vintage newspapers (or any paper products however not people with glossy prints) and cut those into 1 inch thick strips. Soak the paper in water and permit the paper to drip off the excess moisture. Do now not squeeze or wring the papers out, or you may have hard pieces of pulp which the worms cannot use.

If you're planning on the use of corrugated boards or different thick portions of cardboards, it'd be satisfactory to shred these into tiny pieces. Wet those as mentioned above.

Line these soaked strips of paper or cardboard on the lowest of the bug packing containers up to 12 inches in top. Make certain that the whole thing is loosely organized in order that there are masses of areas between the paper strips.

Sprinkle the paper lining with handfuls of loose dust.

Place some portions of chopped up or shredded food scraps on pinnacle of the dirt. Choose those which can be particularly dry, and no longer dripping wet.

As a rule: the greater various your worms' diet is, the higher compost they produce. But that doesn't mean that you may add pretty much some thing in their compost pile. Avoid giving them salt or salty meals objects, as these will kill off a few or most of the worms.

Other items which you need to not contain inside the bug compost are:

All styles of meat products

All kinds of dairy merchandise

Cooking or baking grease and oils

Also, make certain that their diets are strictly kitchen scraps. This method that human and puppy wastes are without a doubt off the worms' menu. Some of the more secure meals gadgets you may encompass in the computer virus bins are:

Coffee grounds and even the paper filters, simply make certain which you shred the paper filters first

Ground up egg shells

Raw or cooked culmination and fruit scraps (i.E. Cores, peels, pits, pulp, membranes, seeds, woody stems) however keep away from the beverages (i.E. Sauce and juices)

Raw or cooked greens and vegetable scraps (i.E. Cores, peels, pulp, seeds, tops, woody stems) however avoid the liquids (i.E. Sauce and juices)

Shredded and moistened paper products (but no longer the smooth types of paper)

Stale bread and different grain-primarily based products (i.E.Cooked oatmeal, popped barley seeds, wheat grains)

Tea leaves and spent tea bags; take away the metallic staples at the tea bags (if any) and shred these earlier than placing inside the compost

Once you've got delivered a small amount of meals into the compost, bury the whole thing with extra handfuls of loose dust.

Allow this combination to face for at 10 mins earlier than introducing the worms to it. This will give the soaked paper time to sink under the burden of the free soil. If there are pieces of paper or meals uncovered at the floor, cover these with soil as nicely.

If you're the usage of a couple of boxes, prepare these as a result.

Finally, just earlier than you area in the worms, sprinkle or spray a light layer of water at the topmost part of the prepared bedding.

4. Divide your worms in step with the wide variety of compost containers you have got organized.

You don't want a precise matter. Just make a visible approximation of dividing the worms similarly the various bins. Place the worms on their beddings. Healthy ones could without delay push themselves down into the pile.

In case you stumble upon some gradual ones, absolutely shine a light over those. They have to wriggle themselves into their beddings eventually.

Tip:

It is unavoidable to have some useless worms in the pile of healthy ones. And you may see these immediately while you first region the worms of their new beddings. Although it could sound counterintuitive, you want to choose the useless ones out and discard well (preferably along with your pets' wastes or spent kitty litters).

If you do not, insects and microbes that feed on lifeless or decaying animals could "invade" your computer virus compost. It is very possibly that you may see greater of those than your actual worms.

five.

Feed your worms simplest a handful of meals in keeping with day.

You don't need to overload your compost with an excessive amount of kitchen scraps that these sooner or later liquefy into mush. The resulting liquid isn't compost tea, but alternatively leachate which the worms avoid as well.

If you do not eliminate the leachate, those might emanate foul odors that entice the undesirable interest of bugs and different smaller pests which might feed at the worms. Also, the leachate would possibly end up poisonous enough to suffocate your crawlies.

If you are not particularly sure how lots you have to feed your worms (in any case "a handful" is not simply a reliable shape of size) simply feed them new scraps while most of the meals objects from the earlier feeding have almost disappeared.

You additionally do now not want to push their meals down into the dust. Just place the meals gadgets on pinnacle of the compost pile. The worms could eventually locate those.

Tip:

After the initial feeding, you don't need to cut up or shred their food anymore. Placing huge chunks of kitchen scraps won't compromise the fitness of the worms. Just make certain which you nonetheless manipulate the quantity and amount of meals you put in to keep away from developing leachate.

6. Control their light.

These worms are photosensitive. It method that they can't stand to be underneath the light for lengthy durations of time. In many instances, overexposure to direct sunlight kills them off. If feasible, save your compost bins in shaded regions, far from direct sunlight or some other mild source (i.E. Near mild bulbs, if you are stacking the bins).

You also can throw a material over the compost packing containers, however make certain that the breathing holes remain open. One of the essential risks of using cloths (or any other cloth) as compost covering is they trap the heat within the containers.

7. Control the packing containers' internal temperature.

Red wiggler worms thrive quality in warm environments that are approximately fifty five°F to seventy seven°F (13°C to 25°C). Anything higher or lower can make the worms sluggish and work (consume) less efficaciously.

8. The worm beddings need to be wet all of the time.

Every morning, or as soon as earlier than bedtime, spray a mist of water on the compost pile. If the climate or weather is especially warm or arid, mist at least two times a day. You ought to do that while the temperature peaks.

Most of the moisture inside the malicious program compost frequently comes from the meals scraps you region in. So you need to take these into account as nicely. The remaining element you want is swamp your worms with too much water or leachate.

Tip:

The worms are suitable signs that their beddings aren't wet enough or are too wet for consolation. If the worms start mountain climbing up the perimeters of the boxes, then it means that their beddings are too dry. They generally try this when there isn't enough meals as nicely.

Mist the compost boxes right now and add greater scraps of food. Gently nudge the "escapees" back into the topsoil. Never mist the worms directly.

If the beddings are too wet, maximum of the worms could drift to the floor, seeking to get as a good deal air as viable. To correct this, you want to:

Gently push the contents of the compost to one facet.

Tilt the box within the opposite direction.

Using antique newspapers, soak the extra fluids or leachate out. Discard those. Don't use these inside the worm compost.

Aerate the whole compost pile via lightly fluffing it up using your hands or with a small plastic digging shovel (with a blunt area).

Push the contents again to 1 aspect and drain off the excess fluids as before.

Take several handfuls of soil and lightly comprise these into the compost pile. Fluff the contents with each addition, then top it off with extra soil. Let this relaxation for 30 minutes. If the worms continue to be underneath the soil, then everything is quality. If not, switch your worms into any other field and start over with developing their bedding.

Next Steps

The cycle of feeding and misting ought to go on for any other 6 months earlier than you may harvest the worms' castings.

Passive Harvesting

Passive harvesting is the process of harvesting the worms from the compost and using the worms to start a new batch of vermicompost. This would depart the newly created compost to be able to use for your vegetation.

This shape of harvesting would probably take among 2 to 4 weeks, and this works particularly nicely when you have several compost bins.

1. Prepare a new compost bin, entire with trojan horse beddings, but no meals scraps. Mist as wanted.

2. In an active bug bin, vicinity all of the food scraps in one corner most effective. Red wiggler worms routinely observe the fragrance of meals. Do this for at the least 3 days earlier than harvesting. Harvest the compost that is farthest far from the meals source. You may additionally discover a few straggler worms here and there, and those you have to placed select up lightly and

region in the new compost bin. Only when there are real worms in the compost can you begin placing portions of meals.

Also, you can find small seed-like eggs in the compost. These are mild brown in colour and seem like grape seeds. These are malicious program cocoons. Gently switch those to the brand new computer virus bin as properly.

3. Continue this sluggish compost harvesting over the span of two to 4 weeks until you have got maximum of the worms out of the antique bins and into the new ones.

You can use the compost substances you have got harvested without delay, or you could dry these out fairly in order that those acquire a crumbly texture.

How to Know When the Compost is Ready to Use

You will know when the compost you are making is prepared to use as it could have a dark, rich look to it and will disintegrate in your hand when you hold or touch it.

Chapter 7: The Proper Aggregate Of Carbon And Nitrogen

All organic be counted includes huge quantities of Carbon and smaller amounts of Nitrogen. The balance of these two elements in an organism is called the Carbon-Nitrogen ratio. In order to obtain the quality bring together, composting microorganisms want the ideal percentage of Carbon and Nitrogen for the synthesis of protein.

Scientist have found out that the fastest manner to produce fertile, candy-smelling compost as a way to reap a Carbon to Nitrogen inside the ratio of 25 to 30 elements, to at least one element Nitrogen. If this C-N ratio is too high, this is excessive Carbon content material, decomposition will simply slow down. On the alternative hand if the C-N ratio is just too low (too much of Nitrogen), your pile may additionally stink so horrific that; you may don't have any desire than to put on a face mask!

Do now not forget about that most materials used for composting certainly do now not have the appropriate C-N ratio of 25-30:1. Hence, you need to do your high-quality to create the best compost recipe possible. You can lessen the Nitrogen content of any compost by way of adding grass manure. To raise a low C-N ratio, you could add paper, dry woods or leaves.

Just try and attain the quality compost viable. You may also even need to deal with excess nitrogen and bear the scent; simply to be sure you've got sufficient Nitrogen to hold the pile "cooking" properly. Hot compost makes a pleasant meal for the ones green leafy creatures!

Estimated Carbon-to-Nitrogen Ratios in various substances are generally referred to as Browns, because of this High Carbon C, to Nitrogen content material N. The following gives examples of C:N ratios in common family substances:

Ashes/wood 25:1, Cardboard, shredded 350:1, Corn stalks seventy five:1, Fruit waste 35:1. Leaves 60:1 Newspaper, shredded one hundred seventy five:1 Peanut shells 35:1 Pine needles eighty:1 Sawdust 325:1 Straw seventy five:1 Wood chips four hundred:1

Greens = High Nitrogen (C:N) Alfalfa 12:1 Clover 23:1 Coffee grounds 20:1 Food waste 20:1 Garden waste 30:1 Grass clippings 20:1 Hay 25:1 Manures 15:1 Seaweed 19:1 Vegetable scraps 25:1 Weeds 30:1

Many ingredients used for composting do now not have the appropriate ratio of 25-30:1. As a end result, maximum should be mixed to create "the right compost recipe." High C:N ratios can be lowered if you add grass clippings or manures. Low C:N ratios may additionally as properly be raised by means of including paper, dry leaves or wood chips.

Breaking Down Particles for Composting

Grinding or shredding uncooked substances is very beneficial, specifically while substances which includes leaves, corn stalks and woody vegetation are used for composting. Shredding will generally disclose a big floor vicinity so that it will make it more vulnerable to invasion via micro organism. However, we can not say the equal component for big portions of wooden or leaves stacked together because, it doesn't decompose without problems in a compost pile. You have to recall the truth that, inadequate supply of Oxygen within the center of a wooden bite or a wad of leaves does no longer necessarily increase the rate of cardio decomposition.

Shredding has been recognised to make a compost fabric extra uniform in length. It has additionally been recognized to aerate the soil, making it easier to address and hold it moist. When compost particles are small sufficient, they emerge as extra calmly heated and may face up to an excessive amount of drying at the floor degree. Once this is done, you are sure to insulate the compost pile in opposition to any warmth loss. In addition, it additionally resists moisture penetration because of rain. You additionally growth your manage over flies while the compost fabric has been shredded or pulverized. It will become smooth to use uniform compost crafted from shredded substances to the land in comparison to one that's no longer sized.

The fine sized composting substances must be less than 2 inches in the most important dimension. However, you could need to also compost larger particles in a manner that satisfies you. If you plan to apply the substances to your flower garden or lawns, then you will want to bypass the compost thru a 1-inch screen to make it appear higher, practice extra without problems, and work into the soil with out a great deal issue.

Sometimes, it may now not be fee-powerful to feature cost and labor in order to shred the cloth. You may also display out, fork or cut up large debris. Bear In mind that, uniformity isn't as crucial for agricultural fields as it's miles for home gardens. Your very first compost shredding isn't important; regardless of the kind of cloth involved. It is simplest recommended which you shred handiest massive pieces of organic materials. Should you stick to the use of massive abnormal portions, it will create large air areas and subsequently, extra Oxygen might be trapped within.

Large, tough feedstock may additionally need to be ground, to be able to increase decomposition rate. You need to never grind up vegetable or herbaceous count due to the fact those materials can emerge as soggy pretty without problems. In aerobic composting, you will find out that high moisture content material is regularly too hard to control. The kind of materials used for composting will provide you with an critical clue approximately when precisely you should shred.

You can also need to grind once more while the composting is mature, or on the tail give up of the maturation procedure. If you choose to regrind on the end of the lively decomposition period, that would be the closing turning before you depart the pile to stabilize. Always undergo in thoughts that shredding /grinding is useful as it will shorten the time for the decomposition system.

How to Manage Moisture and Air

It is important to have moist compost, even though not too wet. It is real that moisture and air amount (aeration) may be discussed separately while composting is discussed, however, the strategies of handling moisture and air are like twin brothers. There are limitless pore spaces which flank the natural debris in the compost cloth. Don't forget about that pores deliver room for air and water to move through the substances of the compost.

When there's no ok supply of moisture, the decomposers will percent up and bid you farewell. On the opposite, if you have too much water flooding the pores, air flow is reduced so you get caught with that stinking anaerobic compost to deal with. Your goal need to be to provide top-rated moisture stability and aeration ranges in an effort to hold suitable conditions for the composting. That way, your effort will now not pass down the drain.

Most organisms that assist to interrupt down the organic remember will want moisture to perform very well. They do their magic in skinny movies of water discovered on organic debris floor. You should in no way allow moisture ranges to drop underneath 40% and drying out of compost materials. If you do, the various organisms which can be beneficial inside the decomposition system may die or be rendered inactive. Forty - 60% by means of weight is the correct of the moisture content material. You really do not need to weigh whatever. All you need do is squeeze out some amounts of materials from the numerous elements of the pile. It's like a sponge that has been wrung out. If this fails, then you could don't have any desire however to add some water. You can also conserve moisture to your open compost pile if you cover it with a tarp.

Soggy substances commonly negatively have an effect on your composting materials. When you have moisture content material exceeding 70%, there is a good danger that you are blockading air waft... And what do you have? A stinking anaerobic state of affairs! In addition, vitamins additionally tend to leach out of compost piles which are completely moist. But how do you already know while the pile is just too moist? A compost pile is considered too wet if you could squeeze out up to 2 drops of water from a handful of the substances.

The usefulness of the pore spaces is to provide essential Oxygen which ensures that organisms thrive nicely. In addition, high-quality pore areas also permit Carbon dioxide, one of the byproducts of decomposition, to escape. When there is enough aeration, excessive temperatures can be maintained, thereby increasing the charge of decomposition. In addition to that, weed seeds and harmful microbes also get killed within the system. If you live in the rainy areas of the world, it's miles really useful that you cowl your compost pile at some stage in the rains. That way, you generally tend to save you it from turning soggy. You need to best-tune your pile's moisture and air tiers. How do you do that?

First, you could turn the natural count so that it will allow in extra air and dry out the moist materials. Your compost pile will now not stink or cross horrific in case you aerate it properly.

Second, you should add materials containing Carbon in an effort to soak up the excess moisture. Materials rich in Carbon consist of leaves, sawdust, and straw and so on.

Third, you have to rewet the compost material on every occasion you watched that it is drying out. It is even most advocated which you rewet the material mainly, while you are turning the pile.

Chapter 8: Why Choose Compost Gardening For Your Garden?

Compost gardening has many benefits. It is a panacea for your property lawn soils and it has additionally some advantages from consistent applications. There are many regions within the world where the soil is not most excellent for any sort of gardening, as an example Felton, CA has a totally sandy soil that firstly leached water and vitamins like a sieve. Years of compost including to this sandy soil has very well improved its capacity to preserve water and nutrients. If the soil of your location is sandy, clay, or fairly in among, then you could use compost gardening to improve the pleasant of your soil and hence make it extra efficient.

Building soil is an important idea in compost gardening. There is a commonplace notion among farmers and gardeners called "feed the soil to feed the plant." In including manure to the dust, you're renewing the shop of natural remember and supplements that are delivered out with the greenhouse merchandise. Crucial to all soil organic groups, herbal remember is the nourishment for soil organic entities. By fertilizing the soil, you are bolstering the dirt animals, from the littlest microbes to the longest worm, which as a result make dietary supplements available for your greenery enclosure flowers.

Some Common Benefits of Composting

Composting has following primary advantages: --

Utilizes waste material: -- Composting uses the waste material of your home in a best manner and convert them into the wealthy organic cloth. Nothing is taken into consideration as waste for nature and the whole thing is food for something else. We can use our herbal assets in a extra powerful way this manner.

Improves soil best: -- Today, the fertilizers, chemical substances, and insecticides that we're the usage of in our gardens and farms are making our land barren. The soil of our farms and gardens is dropping the herbal components that a plant needs for proper growth. We can grow a few flowers in bulk for sometime through using fertilizers, however we ought to soon suffer whilst our soil will become non-efficient. Composting is a pleasant natural way to restore our soil with essential vitamins and minerals without harming the natural functioning of soil. Compost improves the bodily, chemical, and biological properties of soil. This has many advantages like proof against erosion, drought, and stabilization of pH degree. Compost also increases the presence of vital soil microorganisms and suppresses any type of plant disease.

Saves cash and assets: -- Composting makes use of the organic material that you usually throw. You can decrease your garbage bills and utilize the waste fabric for beneficial use.

Composting offers you herbal fertilizer and saves your fee of purchasing high-priced fertilizers for your house lawn or farm. Compost also facilitates to retain the water level within the soil and so potentially reduces your water bills.

Makes gardening less complicated: -- The industrial way of gardening or farming wherein we use fertilizers are very tough to learn and apply. You have to take care of many factors in industrial gardening like the pH level, nutrient ranges, soil temperature, and many more. On the alternative hand, compost gardening is lots easier to apply and study. Compost acts as a soil conditioner, natural fertilizer, mulch, and evidently controls the soil temperature so it makes the protection of soil fitness less difficult.

Use Compost Gardening and Go Green

Compost gardening is the pleasant way to go inexperienced and reduce the extra fee of the use of high priced fertilizers on your lawn. There are many elements of our way of life that we can look at to lessen the quantity of waste we generate. You need to employ the approach of the use of four R's to decrease your personal footprint at the environment referred to as: Reduce, Reuse, Recycle, and Rebuy. Follow the subsequent not unusual techniques for adopting pass inexperienced method: --

Donate and Reuse Your Items

Have you ever found out that how many items you throw away that can be reused? Instead of throwing your reusable items, you may donate them to the individual that needs them. Consider sending your reusable items to the landfill donate your hand me down cloths or much less than new appliances, CD's, furniture, books, gear, and different gadgets that you think may be reused. You can without problems promote some unused gadgets on auction and earn a few dollars from them. Finally, if something left that is bio-degradable, then you could use them for compost gardening.

Reuse Packaging Materials

Paper or plastic baggage, bins, glass or plastic bins, or maybe mailing envelops may be reused if individuals are willing to give minimum greater time it takes to accomplish that. You can without problems send those gadgets for recycling. Limit the usage of plastic bring bags to move green and use your own convey baggage at the same time as going for a buying. Remove any kind of plastic materials from your compost materials because they may be now not bio-degradable and could reason hazard in your plants.

Remove Hazardous Waste

Every household garbage has a few type of hazardous waste substances like syringes, motor oil, pool chemical compounds, fertilizers, weed killers, and batteries. Many batteries and electronic wastes contain lead that is very dangerous on your health, so make an extra bin and placed all of your dangerous substances in that bin for correctly dumping them. Never use those kinds of material in your compost pile and cast off them from your materials straight away. You have to decrease using those unsafe materials to adopt a green life-style. Reduce the quantity times you operate the product. Only buy the amount you need. You can try to use natural products for the motive or much less poisonous merchandise. Donate unused portions to the people in need. So observe those commonplace steps to head inexperienced and stay a healthy life.

A lot of folks who are new to composting will add an activator because they aren't completely sure what they're doing and whether or no longer they need one. Well now you recognize the truth and if you are balancing your compost proper then you can keep yourself some money.

Activators are available a number of one of a kind forms, from powders to beverages and to pellets. Some not unusual activators include blood meal, alfalfa meal, bone meal and so forth. Other activators are based totally on enzymes that, while you mix them with water, come lower back to life and might then work to prompt your compost heap for you.

The cheapest, and possibly simplest to get hold of, is manure from chickens, goats, rabbits or even cows. This can be fresh or dry and you can buy those from most lawn stores.

With any of the food, like bone or blood meal, you can scatter a little on top whenever you add some thing to the pile. If you're including plenty or using an open compost pile then you could have a number of inexperienced and brown layers. In this example you scatter a bit on top of each layer.

Fresh manure can be placed on pinnacle of inexperienced and brown layers. Put a layer of round two or three inches of manure on and with the intention to sincerely kick begin your compost! If you are the usage of dry manure then you want to simply add a dusting now and again or whilst you upload fresh scraps on your compost pile.

The best manure you can use comes from animals that eat grass. This may have the best level of nitrogen in it. Animals such as chickens, rabbits, desires, llamas, sheep, cows and so on are best to accumulate manure from. You are probable to find a person on your neighborhood area who may be more than satisfied to provide you as tons manure from their animals as you could cast off with you!

Horse manure needs to be thoroughly rotted before you use it due to the fact a horse doesn't completely digest the seeds from the grass and plant life it eats and also you do no longer need those on your compost pile.

Whether or not you decide to use an activator is absolutely up to you. If it's far chilly or you're finding it tough to get sufficient nitrogen rich material into your compost bin then it's miles worthwhile using one to kick begin the manner and ensure which you get the first-class compost you need.

Chapter 9: Materials For Your Composter's Garden

Volunteers eat a lot of time discovering this trouble with subscribers as both facets want to create greater compost and much less waste. Ultimately, an awesome blend of materials and conditions is on the heart of any composting method. This bankruptcy covers the corporation of many happy weddings.

Getting to Identify Compostable Materials

We need to begin with the aid of organising a simple reality about compost: When you work with compost, your activity teaches you a way to take care of the fabric.

Starting with rule range for garden compost, "paintings with what you have," it best takes you a few seasons to turn out to be familiar with the right ways to manipulate the maximum considerable components available for your lawn. For maximum composting gardeners, those sources consist of leaf clutter, kitchen waste, lawn components, and grass clippings, so we will spend many pages looking at every of these classes. Dung merits a number of attention too, and we hope you discover the whole lot there is to realize approximately even darker materials inside the Odd Compostable Elements panel.

Depending on anyplace you live & who you already know, exquisite composting substances like awful alfalfa or rabbit dung are to be had without cost from neighborhood sources. Anything nearby, reasonably-priced, & at risk of rot can be an exciting addition to composting initiatives.

Our compost hundreds can simplest be omnivorous from kitchen waste. Add to this loads of garden debris inclusive of pruning perennials, waste annuals, weeds, & vegetable residues. A compost heap can become nearly as unique as a fingerprint due to the fact the elements reflect the particular tastes and conduct of the compost gardener who accrued them. Your composting designs may also differ from the ones of your pals, or they will have not unusual traits based on crop residues & agricultural with the aid of-merchandise that are well-known to your place. But the components to your compost probable might not be the same as the contents of a compost bin or heap in any other part of your united states of america, wherein diverse developing situations require other crops & assist exceptional farms.

Counting Compost Miles

One of the essential ideas of the neighborhood food movement is the importance of preserving tune of your meals miles.

The extra your meals travels and tries to control its assets at the back of you, the more the harm to the environment, as measured by way of greenhouse fuel emissions. These dangerous emissions differ depending on the sort of delivery used higher for air and freight than for sea or rain. Your vehicle is inside the high emission class, so that you will collect several miles of compost each time you leave your yard to fetch compost materials. Compost miles depend while you buy packaged supplements, which includes alfalfa flour, or forestall at the solid to acquire rubbish packing containers complete of manure.

Gathering your Goods: start in your Kitchen

It's a comic story that the American food regimen has 3 primary food agencies: fat, sugar, & meat. It's not joking, however the important reason to get this shaggy dog story is that these are three meals associations that ought to not be covered to your compost. Everything else is truthful play, together with foods which might be low in sugar or fats. As with the entirety, there are differences.

Composting kitchen waste has two major blessings: Less unpleasant materials continue to be for your bin and your compost blessings from a completely extraordinary form of nutrient-rich elements. This variety, in flip, helps an extremely good array of microorganisms, that's specially what you want in a full batch of compost. Kitchen waste is taken into consideration a "green" element that provides nitrogen, so if you add a touch greater "green" inside the form of protein-wealthy cereals or seed meal, you can quick remove quite a few kitchen waste. Alternatively, take an informal method, allowing kitchen garbage to decompose progressively on its very own without generating substantial portions of warmth.

Clean-Out-Kitchen Compost Components

In addition to the normal cuttings of fruits and vegetables, your kitchen can be a source of other treasured fertilizers that typically become inside the trash. Let's be clear: we do now not assist the use of "human meals" that can still be determined as feed for the compost.

But even the humblest property owner will occasionally run out of food that has been by accident moved to the back of a closet or misplaced within the fridge or freezer, & few of us continue to exist with out encountering annoying food moths or food contamination. As you discover the dark corners of your closets, revealing an open show of blended fruits from last year's own family reunion or bread blend that expired at some point of the previous presidential administration, turn this capability mess into a present in your compost heap.

Dry Compost Components

Remnants of cereal, oatmeal & pasta

Stale crackers

Rice

Stale herbs

Wet Compost Components

Tea liquid & Coffee

Leaves

Baking mixes

Spices Expired yeast

Moldy bread

Grits, cornmeal, & corn starch

Tapioca

Bread crumbs

Raisins

Dried fruit

Rancid nuts

Dregs from boxes of beer, juice, or wine

Grounds

Moldy produce Sprouted

Condiments & sauces like salsa, spaghetti sauce & ketchup

Fruits & bread Fruits

Vegetables from expired cans

Withered

Freezer-burned vegetables

Green potatoes, carrots, or onions

The absence of such facts might also suggest that the meals is so vain that it predates this good sized commercial practice. Even if you have a tendency to take expiration dates as a guideline rather than deadlines, if the modern-day date and date in the area is extra than a yr, then the peak has already exceeded. Recycle the packaging if you may, and placed those antique crumbs and old espresso in a compost heap wherein they may be beneficial.

Managing Your Kitchen Compost

There are three essential subjects to consider as you devise the most appropriate plan for composting the kitchen waste:

Site choice

Capturing Kitchen Compost

Animal-Resistant Bins

Site selection

Convenience is the important thing to choosing in which to vicinity your compost in the kitchen because you upload sloppy material to the compost in the kitchen as a minimum every other day & sometimes twice a day. It have to be located as close to the kitchen door as feasible, but you should nonetheless have clean access to it out of your lawn. There need to be a tap near the water as greens are often cut and peeled from the facet of the sink earlier than being placed. If you can not locate area proper out of doors your kitchen door, keep a big plastic bucket with a decent lid to your patio or deck and use it to save kitchen waste quickly. Add the bucket contents to the compost heap for your kitchen every few days, after which rinse the bucket.

Capturing Kitchen Compost

Bottomless plastic composters: Floor-standing enclosed compost containers are a great preference for composting for your kitchen, especially if you have a small patio in which each rectangular inch of area is valued.

Community-primarily based composting programs usually provide such composting laminates at minimum fee, and this provide can not be rejected.

Barrel composters: This also are ideal for kitchen waste disposal as you may speedy rotate barrels to mix new substances with old ones, even as closed barrel composters offer superior animal strength. On the alternative hand, barrel composites are normally pretty expensive.

Buy the version that fits your needs and pocket, or begin growing practical fax from a plastic waste bin with a good lid.

Cylindrical plastic trash can: If the waste container is cylindrical, you can drop the field at the floor in preference to being square or square blend the components inner. Or, if you locate the proper barrel with a facet-folding door, you could use your hammer & nail to drill it air holes and wrap it to mix.

Wire-enclosed compost heap: Two-piece stacks, open or half of-closed, also work well as you may turn stacks the usage of the cloth at the extra superior side to cowl new additions to the pile.

For reasons of visible appeal & animal deterrence, it is continually advisable to shield the composting manner in an open kitchen with welded cord, plastic railing, delivery pallets, or something else that prevents fabric from spreading.

Animal-Resistant Bins: Omnivores, such as dogs, raccoons, lizards, mice, rats, & even from time to time bears, are drawn to kitchen compost. If you have got a problem with any of these animals, it depends in your gardening and composting. With the feasible exception of dogs, animal infection is typically overwhelming. Still, you simply don't want to disturb your community by attracting pests with unsafe compost for your kitchen. These animals find meals sources with their sense of scent, frequently 50 times more potent than yours. When you cowl your kitchen trash with thick layers of shredded leaves, grass clippings, or prepared-made compost, you will likely by no means odor the faint scent of rotting broccoli or grapefruit rind.

Harvesting & Using Kitchen Compost

Commercials for industrial composters are regularly advocated gardeners with a shovel loaded with beautifully organized compost that they were probable eliminated from the compress can also try this, however before taking the image, someone selected all pumpkin seeds, peanut shells, and stickers to cause them to within the compost.

Materials & Supplies

Staple gun

Pencil

Metal cleats (2)

Wood screws (four)

Hand noticed

Screwdriver

Hardware fabric

Scissors

Mesh polyester

Drill or nail

One 6'-long 1×2 pine

Measuring tape

Hammer

Polyester clothesline

Composting Ailing Plants & Weeds

Do the entirety which you produce on your use and delight to your lawn. They surely produce plenty of compostable substances. Good news for composting gardeners, almost anything that comes out of their garden is honest prey to go back to their garden thru compost, including diseased plants and testes.

Most fungi & micro organism that motive plant diseases are significantly laid low with competition for nutrients in the subject of microbial fertilization. Those who aren't starving may be poisoned or fed on by using their movement neighbors. Survival turns into a exquisite effort throughout the microbial modifications as a result of temperature extremes, with high-temperature microbes are the deadliest of all.

Weeds that appear in your garden are clean prey for compost, although they have seeds. Green plant cloth and weeds dispose of vitamins and moisture from the soil just like some other plant to your garden. Weeds which have not started to bloom & do no longer have feasible roots .The buds that help them grow into new weeds may be delivered to any compost paintings, but it is vital to limit any probabilities of hemp seeds.

Composting with Leaves

Leaves vary while it takes to decompose, as they are not all similar—skinny sheets with high calcium content and occasional lignin content. For example, the ones deserted by puppies and birches will rot for the duration of one wintry weather, while thicker okayor magnolia leaves may be saved for 2 years or extra, transitioning from fallen leaves to hummus bites. Some gardeners take into account mixing distinct sorts of leaves right into a compost heap to even out the time distinction & that is in part genuine.

Composting Grass Clippings

We don't want to move to this point as to say that each landscape has one garden due to the fact true lawn renovation is usually a sentence. Most homes require additional water & fertilizer. There are compelling questions about the ecological integrity of grounds in a weather in which the grass stripes do no longer healthy nature's layout. For example, the use of water to maintain a green garden is a waste of precious resources in regions with water scarcity. Even if you are the usage of natural merchandise fertilizers, preserving a properly-groomed lawn will price some time & fertilizer.

On the opposite hand, a properly-groomed garden may be properly maintained on its own, with cautious care. It can dramatically enhance the look of your home. And nothing loses precise grass to play outside. Lawns play an crucial role in landscaping by means of creating a feel of transparency.

Most importantly, a garden allow you to revitalize your lawn. The cut inexperienced grass is utilized in numerous composting tasks. The cut grasses also create a remarkable layer if they may be at the ground very skinny layer. You can use a thicker coat when you follow the garden mattress on sheets of newspaper or cardboard.

Composting Manure

Farmers and gardeners made compost mainly to stretch their compost manure supply. Laying hens persevered when mixed with hay mold or corn stalks rot alongside the threshold. Fertilizer-rich manure, velvety things in an effort to spark the imagination of any plant, similar to composting waste, is amusing for composting gardeners.

Even once in a while fertilizing is instructive.

Witness a miracle that has interested gardeners for heaps of years. Okay, so you are not inquisitive about the usage of fertilizers. Your spiritual ideals preserve handiest human members of the family with animals and observe a veganism eating regimen; you can feel more snug eliminating dung out of your weight-reduction plan composting tasks. This may be carried out for ever and ever the use of protein-rich plant nutrition while a robust source of nitrogen is required.

This may be the reason to your concerns about the use of fertilizers. Okay, because you need to understand what you are becoming yourself into earlier than you compromise to easy up the strong in which buddy takes Ola's horse animal manure consists of many bacteria, from which you may get very sick.

Harvesting & Storing Composted Manure

Most fertilizers are taken into consideration the excellent soil conditioners' medicinal blessings. Still, nicely-composted manure is often appeared as proper fertilizer due to the fact it is rich in three critical plant nutrients, i.E., phosphorus, potassium & nitrogen (P-K-N). It additionally consists of many lines of nutrients in conjunction with acids and different beneficial by means of-products composting technique.

Hay & Straw

Grain growers can harvest hay or straw; both materials are easily available if you want to add them for your composting tasks. Most feed shops promote at least two kinds of hay every ball, and most lawn facilities save straw bales. These are very distinct substances! How supposed for intake animals are grasses and legumes cut and green allow dry within the sun, and then tie in square bundles or wrap in massive spherical bundles.

Handling Hay & Straw

Transporting bales of hay or straw is simple when you have a truck, however make certain to take precautions before transporting the package to the vehicle. The parts that pop out are set up at the rugs, and the handiest manner to get them to the latter is really worth deciding on them manually. Avoid this worry with the aid of wrap each bundle in a complete-sized flat sheet before lifting it a car.

Seeking Straw & Hunting for Hay

There can be many assets of hay and straw in rural areas, but locating a few parcels can be like locating the proverbial needle in them. Straw for the city and us of a gardeners laws supply and demand additionally function right here: a bale of straw is sold for $ three at a nearby feed, the shop can value up to 5 instances as a whole lot bought as a festive fall decoration a garden middle or craft keep.

Farmers' ads: The feed store is a distant offer or a far flung provide memory, depending on where you live. You can locate man or woman farmers promote hay or straw via marketing. Check out the maximum local newspaper. You'll find in the main unfastened weekly newspapers. Explore within the fall neighborhoods for houses with balloons used as decoration. I assume to look at the locals and spot in the event that they have any plans for their put up-Halloween straw.

Hayrides: If there are Halloween rides to your location, take a look at with the operator. What occurs to the straw whilst the automobiles are left for the winter? These can be prepared to present you some balls or promote you at a reduced fee rhythm, mainly if the parcels were given stuck inside the rain.

Ornamental grasses: Don't think about feasible assets of straw for your terrace. Like their grain-developing cousins, they develop ornamental grasses. Straw isn't packaged properly but stems dry up at the pinnacle.

Grass and different decorative grasses are great for composting.

Paper & Cardboard to Compost

Money can grow now not on timber, however paper and cardboard! Small in percentage terms, the file is constituted of short-lived flora such as cotton, but most paper and board start with cellulose. With several except, paper and cardboard can be composted and different tree via-merchandise along with leaves, sticks, sawdust, or wood chips. They are wealthy in carbon (brown) substances and frequently have distinct paper-based materials C / N ratio from 2 hundred/1 to over 500/1. When it is broken or crushed small portions, paper, and cardboard blend without problems with kitchen waste, manure, and different nitrogen-wealthy substances.

Chapter 10: Making Compost With Worms

Vermicomposting is a technique of Composting that makes use of sure species of worms to feed on and convert organic materials into precious soil compost. These worms decompose organic materials into soils pretty rich in vitamins and plant objects called "castings." Castings are simply similar to saying "computer virus poop." In different terms, castings are give up merchandise produced while organic substances have gone thru the digestive procedure in a worm's intestine. It is straightforward, tremendous, silent and profoundly compelling, reworking your kitchen scraps right into a compost that is in most cases steeply-priced to shop for.

Purpose of Using Worms for Compost

Since the worms remodel organic substances into the soil with excellent vitamins, it implies which you don't have to burn cash on getting

fertilizers which might be possibly dangerous for the environment.

Using worms for composting help with getting the air stream, texture and soil structure progressed. Also, moisture is nicely retained. All these will see your vegetation growing stronger, developing roots systems deep into the soil and a excessive degree of resistance to drought.

Worms have been remodeling natural materials into true soils for a protracted time period. These substances are eaten after which decomposed into natural soil wealthy in vitamins, known as casting. In this situation, it's far one true meals source for plants which could serve as a replacement for vitamins acquired immediately from the soil and are lost from plant harvest.

One different large benefit of malicious program farming is that the following compost is brimming with beneficial microorganisms, which the soil frantically wishes. These get natural wastes damaged down within the soil, presenting appropriate nutrients to vegetation.

Getting Started

Worm farming may be as exorbitant or as reasonably-priced as you want it to be. With a chunk of a creative mind and applying recycling approaches, items around can be used, which can be all one wishes to start. What you'll spend on, most possibly, can be the purple worms. However, they may be discovered in rotten varieties of manure, which can be used.

The object you'll require is some thing to house the worms. You can purchase computer virus farming packs in which containers can be discovered or probably build one, which isn't difficult. A lot of trojan horse canisters are either darkish or stupid green on account that worms don't take care of direct daylight. In the event that you buy a bin of clear plastic, do cover or give it a few correct painting on the frame just to fend the light off. A bug container may be produced using timber, plastic, steel, or Styrofoam. These have to, however, be capable of maintain moisture and offers darkness to the worms. It can then feature as a worm bin.

Purchasing Your Worms

You can buy your worms on the internet, from a community company or on the other hand, inside the occasion which you recognize someone who is into vermicomposting, worms may be gotten from them. You will require a pound of worms or possibly 1/2. This depends on the size of your malicious program farm. Worms reproduce in a quick manner, and the populace will get to increase. Ending up with an immoderate range of worms offers you the choice of selling some off to be used as bait, discharge them into the wild or make any other bug ranch. They don't need to be wasted.

Using a Variety of Worms

You can employ a single variety or species of worm to your farm. It is claimed by means of a few that making use of two or 3 exclusive species or kinds can be useful as to increasing yield, and also primarily based at the truth that different species of worms need a various degree of temperature. This way, it gained't be hard to control the temperature.

What goes on in a bug box?

Using a worm bin comes with the aim of installing your waste and getting vermicomposting because the end product in this way, reusing the vitamins.

Worms are delivered as the number one level customers in a everyday bin device. This is in an try to debilitate the fast improvement of mesophilic microorganisms that release high temperatures, which are hazardous to worms. Make an try to maintain an equilibrium of the quantity of meals and worms positioned into the bin. This is executed by means of inputting a meals amount the worms can devour.

Most bug containers have an improved potential of heating up. This is the reason certain meals types that may bring about the creation of outrageous warmness are counseled against.

Naturally, worms won't be visible in compost till the thermophilicmicroorganisms are completed. The remaining microorganisms as well as debris if natural substances turn out to be

what the worms can use as food. Obviously, extraordinary lifestyles forms, like mold, parasites, springtails, grubs, and different bacteria, take part in consuming this tremendous stockpile of meals. A element of these bug growers inculcate pet posts but, many are minuscule and go unrecognized.

For that of a closed bug bin, this technique is repeated till rich vermicomposting is attained and can be applied to farms/gardens. For that that's outdoor, it's miles a device of some thing eats some other, making the heap will have lots greater consumer degrees.

What size container do you require?

To select how giant a computer virus bin need to be, there's a need to first decide the quantity of waste and what kind to be composted. The type of waste can be just kitchen waste or with the inclusion of scraps from the backyard.

Where to Put Worm Bin

When settling on a gap to put a computer virus receptacle, get your work completed regarding

the residing necessities of the species of trojan horse you'll be dealing with. This is particularly basic in case you select the malicious program bin should continue to be outside. In chillier environments, worm bins positioned outdoor have to be blanketed in the less warm time of 12 months, and in very hot environments. The climate shape should be the figuring out element on in which the bin should be positioned, both in a shed, the basement or on a porch. Most worms work feature optimally at temperatures somewhere within the range of 59 and 77°F (15 to 25°C). Worm cultivators concur that temperatures beneath 50°F or above 86°F can be risky for worms.

Bedding for worms

Beddings may be absorbent fabric that is high in carbon which is applied to make their residing space. Over the long term, the fabric could be transformed into castings through the worms, however at a slower fee than their actual meals. Beddings which can be commonplace are shredded cardboard or newspaper. Other super substances which can be used are rotten

manure, peat moss, elderly compost.

Maintain, Harvest, Store your Worm Castings

Most organisms want first rate feeds, and worms are not any exemption. A balanced food plan containing carbohydrates (cellulose), fats, protein, and minerals are on the whole preferred. This consuming routine sounds similar to ours. The organic materials given as food to worms are known as "feed stock" with the aid of cultivators. It is generally a nitrogen-wealthy fabric that likewise gives power to the microorganisms inside the box.

Feeding is one appropriate part of renovation. Vegetables, end result, espresso grounds, kelp meal, plain bread, rice, cornmeal, pasta, are accurate feeds for your worms. Some meals to avoid are meats, citrus, bones, garlic, dairy products like milk, butter, eggs, yogurt, highly spiced meals.

With time, the castings at the buttons will become compacted, with out the presence of worms. This makes it tremendously easy to cope with. Your worms will relocate out of the castings and move up the layers in which there is food and bedding. This is appropriate for single bug packing containers. After migration of the worms to that region, you may simply take it out and vicinity it in a transient bin at the same time as you void your malicious program farm out.

After emptying your computer virus farm contents onto a protecting or tarpaulin, the castings are then separated into numerous cone molded thousands. The worms will keep on retreating to the decrease a part of the heap, meaning you could get to take out the pinnacle layers easily. Continue to take out the pinnacle layer, keeping up ten or fifteen mins then, commencing any other. The worms will hold to withdraw down to the lower a part of the hundreds. At the point when you get right down to the real lower a part of the heap, you can go back that in your bug receptacle to kick off another compost technique.

After harvest has been completed, add new bedding to the bin and add the worms to

kick begin vermicomposting yet again

Chapter 11: Type Of Composting – Tumblers, Boxes, Piles

When finding out what you're going to put your compost in, there are a number of options available from easy outside piles at the ground to enclosed bins and tumblers. While they all have their benefits and drawbacks (speed of composting, neatness, moisture retention and many others.), there are some of points to take into account earlier than choosing the high-quality one for you:

How a great deal waste will you be composting?

Are you able to turning and combining your compost?

How a whole lot space have you to place your compost?

One of the principle concerns may be based totally on how a lot waste you plan on composting. If you have a lawn that produces a massive quantity of waste over the course of a year e.G. Grass cuttings, hedge clippings then you'll want a large device to method those, which include a huge pile or pile gadget. If your composting is especially modest, and you may no longer have sufficient cloth to form a 3 foot through 3 foot with the aid of 3 foot pile, then a smaller composting bin or tumbler may be pleasant for you.

One of the primary elements in composting is aeration and regular mixing of your compost to permit air to enter. To make certain the fastest composting this turning must be performed on a everyday basis. If you are unable to show your compost, as an instance it can be too bodily stressful to move and agitate the pile using a garden fork, you then should remember using a tumbler or different bin which permits for an smooth blending.

A final consideration is the quantity of area you need to make your compost. A compost pile won't be the prettiest item to your garden, though of direction this can be hidden out of view, say at the back of a shed, or will have a perimeter around it (which additionally serves to enclose the pile and potentially conserve warmth and moisture). If this isn't an alternative for you, then it is able to make greater experience to have a bin or tumbler for your composting needs.

Types of composters

A stationary bin might be the most effective and simplest manner for a person with little composting wishes. This is ideal for non-stop, in preference to batch composting, where sparkling fabric is added constantly. You will see a extensive sort of boxes on the market on-line, normally manufactured from steel or plastic, with air vents alongside the facet. This gives an enclosed location in your compost, with the venting bearing in mind entry of air into the compost. To blend the compost, you can virtually steady the lid on the bin and roll it alongside the floor to agitate and mix up the contents. Alternatively you can have a 2d bin and shovel the fabric from one bin to any other. These are especially simple to make your self. To do so, get a 32 gallon trash can, or large, (steel or plastic) but make sure it has a comfortable becoming lid. To allow aeration, drill a sequence of holes everywhere in the facets and base (drilling about 6-12 inches apart). Cover those holes with a cord mesh (to prevent spillage) and away you go. To improve aeration you may region the bin status on some blocks, raising it off the floor. Alternatively you could noticed off the bottom of the bin and location it on uncovered soil (this will improve the speed and efficiency of your composting).

A tumbler bin is similar to that described above, except the cylindrical bin will be set upon a stand to allow for easy rotation. This will normally value more than a simple trash can composter.

Both allow for a neat, compact composting, and must you by chance throw some food scraps into your compost will save you any vermin from coming into your compost. Being enclosed they guard your compost from the elements, and allow for properly moisture retention, saving you from having to again and again water your pile and also can assist maintain temperature within the pile. These are however of limited capacity, even though if you aren't planning on composting massive amounts are perfect for the gardeners with restricted area or balconies.

The 2d sort of composting is a semi-enclosed bin. These, not like those described above, could be semi-dependent (from twine mesh presenting a barrier to brick enclosures) with varying degrees of safety/openness to the elements. These have the gain of retaining a better shape in your composting pile. By having at the least one wall, this will greatly reduce the amount of moisture lost to the environment. A quantity are defined below:

The handiest semi-enclosed bin is one crafted from twine mesh, rolled to shape a cylinder that's then pegged into the ground. To construct such an enclosure, get a 12 foot (three.65m) duration of medium strength twine mesh. Form right into a cylinder, tying together the usage of ties. Simply region at the ground, or restoration to the uncovered ground the usage of a few small hooks. Being fabricated from mesh, this manifestly allows exceptional aeration into the pile, however conversely can allow for moisture loss. To reduce this you can wrap the outdoors and the top of the cylinder with some plastic.

A more state-of-the-art bin may be made via forming an enclosure to keep your compost pile. These can be crafted from some of materials: bricks, timber, pallets, straw bales etc. If using timber or pallets, ensure that the wood hasn't been treated with any chemical substances which can leach into your compost and soil. In all cases you ought to make sure which you construct your enclosure to keep sufficient compost, remembering that less than 3-four toes (ninety-120cm) high, extensive and deep will not have enough extent to have an powerful composting, whilst large than this size will cause compacting of the fabric and anaerobic situations to set in. Also ensure that there's an smooth commencing (or indeed just have three facets) to allow get entry to to turn and cargo the pile. In all instances use a base layer where feasible, of difficult branches to permit for extra aeration. A final type is a multi-compartmented bin, manufactured from 2 or three adjoining packing containers. These are frequently used for heavy users, wherein fresh, maturing and completely prepared compost may be stored one by one, or having one in use always, mixing the cloth from one bin to another while blending. Not be counted what type of enclosed/semi-enclosed bin you use, you may similarly resource the composting by insulating the bin e.G. Covering with old carpet, tarpaulin, having a Styrofoam or straw bale shell.

While I even have defined the way to construct a lot of these sorts of bin, those are of path to be had out of your local shop or on-line, ought to construction be a step too some distance for you! If of course, this all looks like an excessive amount of work or rate, don't forget about that a easy pile of compost on the ground will still produce tremendous compost if tended with a bit care and interest.

Compost Tea

While compost may be used for plenty functions within the soil to improve texture, moisture retention and many others. It may also be used as a low-power fertilizer/liquid feed on your plants and veggies to give them a wonderful herbal raise. To try this, we use finished compost to make compost tea. When we feed nutrients to our flora inside the form of powdered fertilizers they must be solubilized within the soil to a liquid form that the plant roots can absorb. Rather than expecting rain water to leach the vitamins from the compost in the soil, the technique of making compost tea permits this to arise rapidly over a few days, to launch a nutrient and useful micro organism-wealthy solution. This compost tea can be used as a liquid feed to the roots of the plants. Unlike commercial fertilizers this can no longer purpose root burn or harm the plant. As many useful micro organism are launched into the compost tea, this makes a noticeably powerful anti-fungal solution whilst sprayed on leaves and seedlings, so not handiest does it enhance your plants growth, it also can guard them from sickness.

Making compost tea is fantastically easy, with very little equipment needed. Take a bucket or bin and shovel in approximately eight-12 pounds of mature compost (candy smelling, earthy compost) i.E. Compost this is fully decomposed and is prepared for use within the lawn, according to 10 gallons of water (again the amounts and ratios aren't written in stone). Every day deliver the answer a terrific stir with a stick – this aerates and mixes the answer, stopping it selecting the lowest. After five days or so, the water will have modified right into a darkish, tea-shade with an earthy, candy odor – if it smells foul or in any other case, do now not use. Using a cheesecloth as a filter, strain the tea into a brand new bucket. Any similar fabric may be used (material, burlap etc.). The compost this is left behind can be added lower back in your compost pile or definitely used in the lawn. The tea must be used incredibly quickly after guidance (this could't be stored, nor have to or not it's diluted) and can be sprayed onto leaves or seedlings as an anti-fungal solution, or may be used as a liquid feed to offer your flowers a lift. Other strategies of creating compost tea are kind of comparable. Using hosing and a pump you could place the hose at the lowest of your bucket and continuously bubble air into the solution. If you have the device, then that is a best way to make tea, but certainly given the facile nature of other techniques, your cash might be quality spent someplace else. Alternatively, rather than straining the compost at the end, you may make a "compost tea bag" by using shoveling the compost onto a few burlap and with some rope tying this as much as make a bag that's then lowered into the water. Make positive the length of rope is greater-lengthy and hangs out over the brink of the bucket, so that you can without problems dip and dunk the tea bag each day.

If you've got made compost, do that – you'll be pleasantly surprised on the increase the tea will deliver your plants, fruit and greens.

Chapter 12: Composting Tips And Other Concerns

Having a tough time? Although the commands make it seem as although composting is straightforward, you could't really assume people to get it proper immediately. For the ones having a few troubles with their composting, here are a few additional tips that simply might be capable of help:

Temperature

Compost breaks down quicker whilst uncovered to heat. For this cause, it's ideal to hold the compost under the solar, preferably with a temperature between a hundred and twenty and one hundred sixty degrees Fahrenheit.

Compost Size

Chop the substances up into smaller portions. This makes it less complicated for the compost to warmth up, consequently dashing up the method. Make certain to pile on a huge bulk of compost on a ordinary foundation. For instance, attempt gathering all the meals waste for one week earlier than throwing it all out inside the compost heap. This will help with the keeping and buildup of warmth, permitting compost to interrupt down quicker.

Indoor Composting

Indoor composting is likewise feasible, in particular if you don't have the land region vital. When it involves indoor composting however, it's especially really helpful to maintain your compost covered to save you flies. The smell can without problems entice them, and might make it difficult to get them out of the residence.

Rules and Regulations

Ask you local homeowner's association or the town authorities about any rules and rules regarding composting. There might be some regulations regarding bin placement or digging a pit to your backyard. Although there's no question that that is an environment friendly practice, government may also have specific instructions on how it must be accomplished.

Newspapers

Although newspapers and magazines are biodegradable, including them to the heap is not constantly a terrific concept. This is due to the fact newspapers have ink which slows down the breakdown process. It would be better for the environment to have these gadgets recycled.

Troubleshooting your Compost

No remember how easy it'd appear, you may't absolutely count on to get it right the primary time. Some composters may also observe that their compost isn't producing the favored consequences. If this is the case, it is probably a terrific idea to take a very good observe your pile and pay attention to a few problems. Here are some of the most commonplace issues first-time composters tackle:

Bad Smell – a bad smell is a sign of too much nitrogen within the pile. If that is the case, start including extra carbon materials to the pile. If the scent doesn't scent of ammonia however, this usually way that the pile is overwatered. To restore this hassle, attempt turning the compost greater frequently while adding substances that can help absorb the water. This ought to allow the air to bypass through the device, removing the horrific odor.

Pile temperature is past a hundred and sixty Fahrenheit – note that 160 tiers Fahrenheit is the most amount of temperature that the compost have to be uncovered to. If the temperature Is going past that, there should be too much nitrogen within the mix and carbon ought to be included to level the ratio. Aeration is likewise any other appropriate answer.

There are still a few items not decomposed – if some gadgets nevertheless haven't broken down after some time, it's possibly that they're too large. Take them out of the mix, and shred them before mixing it once more.

Rats and different rodents – you may have brought some thing edible inside the compost, together with meat and fish. Animals can most effective get to it if you have a freestanding pile. Either you remove the beef or transfer your compost to an animal-secure bin.

The temperature does now not meet necessities – if that is the case, you won't be adding up enough nitrogen to the system. Make sure that the ratio is well met and growth the pile to generate more warmth.

Chapter 13: Composting Tips And Tricks

Grass clippings are splendid for compost packing containers because they upload more nitrogen that your compost will want. If you don't have any grass due to the fact you stay within the town, ask your buddies with lawns or ask the local landscaper for a number of their grass clippings.

You can't "just" use grass due to the fact all it will do is pile up and stink. No compost. So, similarly to grass clippings, your compost bin may even need masses of carbon so don't forget about the "brown" materials.

They upload a few first rate decomposition and make your compost-wealthy. Just watch out for garden clippings which have been treated with distinctive herbicides and pesticides.

Be certain to ask your friends or the commercial enterprise which you're taking the clippings from. If you're selecting up dead plant life or plant clippings, make sure to ask about pesticides when it comes to those as well.

Adding newspaper or some white paper this is plain (now not the glossy mag stuff) is likewise top in your compost bin. Just shred it first, to help kick begin and accelerate the composting technique.

So how do you understand while your compost is ready? It must sense, odor, and look like dark, wealthy soil. In other words, you shouldn't be able to select out the various things which you've positioned into your compost bin. The contents of the compost bin will also be about half of as huge (in extent) but experience heavier and greater dense.

If you're including worms into your compost bin, worms love used espresso grounds. If you're not a large espresso drinker, many espresso power-thrus, cafes, and different places that serve espresso supply their used coffee grounds to buyers that ask for them. Tomatoes also are massive fanatics of espresso grounds so when you have a bit more, blend it in with the soil around your tomato plant life.

If you're a massive fan of seafood or if you're going to head fishing, you can upload seaweed and algae for your compost bin as well. Make sure that it's now not toxic to you or to different plant life. Also, make sure to rinse off the salt before you add them to the bin.

In the wintry weather, you can preserve your compost pile in a black, plastic barrel and region it in as much direct sunlight as you may do. You can also add a few hay bales with a view to insulate your compost pile from the cold.

If you're working with a trojan horse bin, you're going to want the temperature to stay among fifty-5 and seventy-seven tiers Fahrenheit. You're also going to want your compost to be fairly compact.

The fine place for them would be in a temperature controlled room or in a room that stays the same in temperature for lengthy periods of time, like a garage, basement, or even in cabinets in your house (like under the kitchen sink).

If you need greater carbon for your pile, flip to straw. Straw is a excellent source of carbon. Just ensure which you kind via it so you don't find any seeds from weeds. This will make sure that the tablets are nicely "cooking".

If you're finding your compost pile to be as a substitute cumbersome or difficult to manipulate, take a very good take a look at how plenty you've got. A bin this is complete of "fresh" compost material must be about 3 ft cubed.

This is a good length for mixing and turning the bin as well as allowing properly airflow thru it. If it's far any larger, it can be difficult to turn and the materials in the middle and bottom of the pile received't get an appropriate amount of airflow if you want to "prepare dinner" nicely. If you want more help aerating your pile, use a compost turner every couple weeks.

A first rate way to hold your compost pile while not having to expire on your bin after every meal is to hold a compost pail on your kitchen. Lift the lid up and drop to your desk scraps in, then near it up. You can take it to the bin after it's miles complete.

This will save you some journeys out to the compost bin. Specialized compost pails are fitted with carbon filters to help cast off the odors from food scraps. They also are available in classy and stylish designs that could suit any kitchen or domestic.

Make sure that your compost piles aren't soaking wet. They have to be damp, but. As you build the piles, moisten every layer consequently. How damp?

Take a sponge, soak it in water, then wring it out well. Your compost pile should be as damp because the sponge after it has been wrung out. Keep this degree of moisture in the course of the summer and/or the most up to date months.

If you need greater help with starting a compost pile, a terrific recipe would be, initially, compost starter, cottonseed meal, blood meal, alfalfa meal, and aged manure. All of those are full of nitrogen and the microbes that could destroy down organic matter into the compost that you want for your garden.

If you have got organic depend that breaks down slowly (like corn cobs or stalks that are woody), damage them up with a rock or a hammer to make it less complicated for the microbes to break them down.

Worms can fend for themselves in the compost pile for a pair weeks. So in case you need to head on excursion, you don't ought to fear approximately leaving your compost bin if you feed them before you leave. However, if you're going to be gone for extra than 3 weeks, having a pal stop by way of to check on them and feed them is a great plan.

When it's time to deliver your compost to your garden, do so everywhere from to 4 weeks earlier than you plant any vegetation or vegetation. This will permit the compost time to combine in with the soil and stabilize before you plant. The ph levels and other vitamins that work themselves within the soil need to stabilize earlier than you put your seeds or flowers in.

The ratio that you need to your compost is carbon to nitrogen, thirty to at least one. This is usually recommended so one can hold pests and smell at bay. However, if you spoil it down extra (twenty to 1 or maybe ten to one, carbon to nitrogen ratio), the decomposition will paintings faster.

It all relies upon in your surroundings and your capacity to maintain pests away and odor down. Your buddies won't be a huge fan of you or your compost bin if it smells.

Adding Compost To Your Garden

When you plant: that is the best time to feature compost. Laying your seeds or planting your new plants right into a lawn with lush, compost filled soil is a splendid way to welcome in a new or refreshed lawn.

It doesn't be counted if you're planting plant life (perennials or annuals), trees and shrubbery, or end result and vegetables. Compost will assist all of those things develop vivid and tall.

Incorporating compost into approximately four inches of soil earlier than you plant is a exquisite start but in case you've moved beyond that and have already started planting (and don't want to disturb too much of your lawn), dig your hole and positioned a rounded shovel complete of compost into the hole before you put the plant in.

If you're using compost for larger flora (like shrubbery or timber), use a 50/50 blend of compost and soil. You can do that via digging the hollow, blending what you've dug out with compost, then setting it lower back in: do not forget, 50/50.

Making Your Own Potting Soil

It's smooth to make your own potting soil with only a handful of components. Your potting soil will include 4 components top soil, 4 parts compost, and one component sand. Mix it all up and you'll have the appropriate combination for potted planters, boxes, flower bins, and hanging baskets.

It's Your Pet

Think of your compost as a pet. Compost doesn't do nicely if you forget it; it'll get stinky and disgusting otherwise. You'll need to feed it on a everyday basis, flip it each couple of weeks, take a look at the moisture content each day, and ensure that the "food" you are giving it is healthful and the right balance.

Things You Shouldn't Compost

Organic plant-based totally products need to be the best matters which you positioned into your compost bin. Do now not compost eggs, fish meat, dairy, oily ingredients, canine or cat waste, bones, diseased flowers, weeds and seeds of weeds (then you definitely'll simply be planting weeds into your garden when you add the compost for your flora), and things with pesticides or herbicides on them.

Fast Compost

Composting will take everywhere from three months to about six months, depending on how a lot you are making. However, you could try to make compost quicker by the use of the "Berkley" method. Here are the Berkley suggestions:

Shred all of your compost substances (by and large the carbon or dry additives). You must additionally try and shred nitrogen or green components as lots as you may. The moisture content will generally make this greater hard.

Use a shredder, pruning sheers, or a chipper to assist.

Get the ratio perfect: 30 carbon to 1 nitrogen.

While it still has no longer been confirmed via scientists, a few people clearly trust that the Berkley approach is quicker and extra powerful. It does require a piece greater effort however ultimately, it may be well worth it for you, depending to your stop goal and timeline.

Chapter 14: Gardening Hints For Beginners: 10 Simple Steps To Breakthrough

#1 Gardening Tip: You must build on a Good Foundation.

First of all, you need to verify the carbon and nitrogen ranges on your soil. You want to begin with a stable foundation on your spring gardening length, and those pointers will serve. If you've got very poor and unsatisfactory soil, I deeply endorse that you begin with a square foot lawn. You can either purchase plastic ones from gardeners or make a wooden container, which was what I did. You can then add clean topsoil from the store and upload composted and mulched leaves. This is the most reliable manner to begin.

#2 Gardening Tip: You Must Ensure Good Drainage.

When you have installation your rectangular foot garden, you ought to make certain that the drainage is excessive-grade, otherwise, your vegetation will no longer final and your gardening vocation could be nugatory. Be positive that you supply simply as a great deal water because the plant demands, and you'll realize this quantity through your gardening adventure.

#three Gardening Tip: You Must Provide Lots of Sunlight.

Are you conscious all fruitful vegetation is dependent on the solar? Without sufficient sunlight, the garden will by no means be effective and possibly never germinate. This is important. Nevertheless, you must make certain that the flowers that are in the solar can bear it. You will locate all this statistics is on the back of every seed packet. The rectangular foot lawn region have to be accurately located in keeping with sunlight and drainage area.

#4 Gardening Tip: You Must Make Variety.

For a flourishing lawn, you must add quite a few range to your garden. This will help with soil exceptional and pests. If you vary the various plant life, the bugs will generally tend to now not come lower back due to the insurability of what will be in the garden. Furthermore, it is typically regarded that all farmers rotate their crops for higher soil. As a gardener, you need to do simply the identical!

#five Gardening Tip: You Must Plant on the Proper Time.

This might also appear clear, however it's far crucial. If you plant the seedlings too soon, they will be frozen by means of the frost of the early season and all the work can be for not anything. If too delayed, they may now not be capable of acquire intensity earlier than the solar beats down on them and shrinks them, or a fall frost kills them. Always investigate the seed packets for this statistics.

#6 Gardening Tip: You can Jump Start Transplants.

Make hopefully sure which you purchase fresh and wholesome transplants. It is tons greater practical to pay greater coins for first-class flora as a way to return a hundredfold in return in your work. Make certain as well to offer the transplants a whole lot of water, or they will not live to tell the tale to flourish. Nevertheless, the developing season is not as long as several human beings might also trust, so you must get the flora into the ground on the normal time, and they will hit the ground spreading!

#7 Gardening Tip: You Must Oppose Over-planting.

As a beginning gardener, it's miles most beneficial to just begin modestly. As you perfect all the techniques of harvesting and developing numerous styles of merchandise, you'll then be capable of set up greater of what you preference.

#eight Gardening Tip: You Must Present Lots of Nutrients.

After you have began the seeds and they're flourishing, you should make certain which you proceed to give them vitamins. Depending at the flora you are developing, blood meal and fish emulsion are immeasurable items to feature to the soil. Besides, all kinds of compost consisting of kitchen scraps, leaves, and different nitrogen-rich composts will make your garden thrive. Of path, understand the rule of balance.

#9 Gardening Tip: You Must Discover Pests Early.

As your flora are becoming larger and higher, continually check out for the symptoms of pests at the plant. Are the leaves chewed and holed? Is the plant drooping because of some sickness? In that case, like a certified gardener, you have to take the moves to defend your plants. If you are being bombarded by using rodents and deer, you could positioned a radio or net out near the lawn to misinform and frighten the pests, and blood-meal will stop the deer. As for all diseases, there are organic recipes at the marketplace which could help with them.

#10 Gardening Tip: Prevent Weeds Before They Begin.

This is important. It is critical while gardening the old style route due to the fact I did no longer deal with weeds and they took over the lawn right away. Just a brief care each day saves mins and days of tough and additional exertions. Also, a herbal herbicide of vinegar or some straightforward repression of corn gluten meal can also lighten your mission of casting off the weeds.

I agree with those gardening guidelines helped! I carried out them in my gardening as properly. Make this effective gardening recommendation ever be part of your gardening arsenal!

I wish these gardening hints helped! I could be imposing them in my gardening as nicely. May those gardening hints ever be a part of your gardening arsenal!

Conclusion

Making your own compost is remarkably smooth to do and is sincerely top amusing. It is a completely environmentally pleasant factor to do due to the fact you're reusing waste that would otherwise ought to be eliminated by way of the municipal trash organisation for reprocessing, recycling or incineration.

By composting you lessen the quantity of waste that wishes to be removed from your own home. You additionally lessen your need to purchase compost, a number of which is harvested in a totally environmentally unfriendly way. A lot of the compost you buy is made in exactly the equal manner that you might make it at home, simply on a much large, commercial scale.

Often when you purchase compost you will find sticks and different things in it, which a person has placed out for the garbage collection and it hasn't damaged down at some stage in the composting technique. This indicates you why you need to make certain which you do no longer positioned thick stalks, fruit stones or hard, fleshy stems in in your compost.

You will want to have somewhere to position your compost and permit the biological system of composting to take vicinity. This may be a huge plastic bin or a wood container. It relies upon upon your finances and personal preference as to that you go for.

Then you acquire kitchen scraps and garden waste including weeds (though no longer perennial weeds), grass cuttings, leaves and so forth. These are left in your compost heap and bacteria, worms and different matters break it down into humus, which you may use as compost.

If your compost bin lets in it, you may need to turn your compost often to aerate it and to assist the bacteria to interrupt down the material. However, a few compost packing containers and the big plastic Dalek formed containers specially do not truly lend themselves to turning the compost effortlessly. You can purchase a compost bin that is on a spindle, which you could rotate to show your compost over, almost resultseasily.

You can pace alongside your composting by using paying special interest to the airflow and to the temperature of your compost heap. For maximum human beings that is manner an excessive amount of work, so if you construct your compost heap to consider those desires, you then gained't need to fear approximately it a lot. At the give up of the day, in case you pile up your waste anywhere, nature will paintings its magic and destroy it down into fertile humus. Composting is all approximately taking this herbal process and supporting it alongside and making it quicker.

Making compost is a quite honestly system and you could enjoy making your personal at domestic. There are plenty of benefits from having your very own compost at domestic, now not least of that is the discount in waste produced by your household. You recognise precisely what is going into your selfmade compost and you understand it is normally free of harmful chemical compounds or insecticides.

Anyone can make their personal compost at domestic; whether or not you have a large or small lawn you may get yourself a compost bin, or boxes, and begin composting nowadays. Compost bins are readily to be had on line and from local garden supply stores. Some neighborhood city councils are running to inspire composting (because it reduces their expenses by using decreasing the quantity of trash to be removed) and will promote you a compost bin for a superb fee.

It isn't very luxurious in any respect as a way to begin composting and while you do (and also you start the usage of your very own brilliant humus on your lawn) you may be addicted to composting and like to do it. You'll locate yourself persuading pals and circle of relatives participants to begin composting and appearance to amplify your operations as you comprehend how much you may truely compost and re-use in your property and lawn.